INNOVATIVE AND INTELLIGENT TECHNOLOGY-BASED SERVICES FOR SMART ENVIRONMENTS – SMART SENSING AND ARTIFICIAL INTELLIGENCE

PROCEEDINGS OF THE 2ND INTERNATIONAL CONFERENCE ON SMART INNOVATION, ERGONOMICS AND APPLIED HUMAN FACTORS (SEAHF'20), HELD ONLINE, 14–15 NOVEMBER 2020

Innovative and Intelligent Technology-Based Services for Smart Environments – Smart Sensing and Artificial Intelligence

Edited by

Sami Ben Slama
King Abdulaziz University, Saudi Arabia

Fethi Choubani
Innov'Com Lab, SupCom, El Ghazala Technopole, Tunisia

Cesar Benavente-Peces
Universidad Politécnica de Madrid, Spain

Afef Abdelkarim
National Engineering School of Carthage, Tunisia

CRC Press
Taylor & Francis Group
Boca Raton London New York Leiden

CRC Press is an imprint of the
Taylor & Francis Group, an **informa** business

A BALKEMA BOOK

CRC Press/Balkema is an imprint of the Taylor & Francis Group, an informa business

© 2021 Taylor & Francis Group, London, UK

Typeset by MPS Limited, Chennai, India

Library of Congress Cataloging-in-Publication Data
Applied for

Published by: CRC Press/Balkema
              Schipholweg 107C, 2316 XC Leiden, The Netherlands
              e-mail: Pub.NL@taylorandfrancis.com
              www.routledge.com – www.taylorandfrancis.com

ISBN: 978-1-032-02030-3 (Hbk)
ISBN: 978-1-032-02031-0 (Pbk)
ISBN: 978-1-003-18154-5 (eBook)
DOI: 10.1201/9781003181545

Innovative and Intelligent Technology-Based Services for Smart Environments – Ben Slama et al (eds)
© 2021 Taylor & Francis Group, London, ISBN 978-1-032-02030-3

Table of contents

Section IV: Digital marketing, smart tourism

Innovative and Intelligent Technology-Based Services for Smart Environments – Ben Slama et al (eds)
© 2021 Taylor & Francis Group, London, ISBN 978-1-032-02030-3

Preface

We are honored to bring you this collection of papers from the 2nd International Conference on Smart Innovation, Ergonomics and Applied Human Factors (SEAHF'20) which was held in Hammamet, Tunisia from 14th to 15th November 2020.

The primary goal and feature of this conference is to bring together academic scientists, engineers and industry researchers for knowledge exchanging and sharing their experiences and research results in various areas of Smart innovation, digital Smart as well as energy management.

SEAHF focuses on original research and practice-driven applications encouraging authors to extend and improve their research activity seeking smart and innovative ideas while considering human factors as a road map to reach excellence in research-development-innovation. SEAHF is aimed at providing a common link between a vibrant scientific and research community and industry professionals by offering a clear view of current real-life problems and how new challenges are faced, supported by information and communication technologies. SEAHF offers a balance between innovative industrial approaches and original research work while not forgetting relevant issues affecting citizens: individual and collective rights, security, pollution, health, applications and new technologies.

Real life issues are the goal of this conference including industrial activity, energy generation, education, business and health. In order to cover such relevant areas in human life, the conference is organized in four sessions: Smart Sensing and Artificial Intelligence (S2AI), Green Energy Production and Transfer Systems (GETS), Telecommunications and Computing Technologies (CTCT), and Digital Marketing, Smart Tourism (DMST).

In order to guarantee the quality of the accepted and published papers several committees have been created: Associated Editors Committee, Advisory Committee, Organizing Committee, International Program Committee, Local Committee, and Conference Support Members. The program consists of invited sessions and discussions with eminent speakers covering a wide range of topics in each session. This rich program provides all attendees with the opportunities to meet and interact with one another. We hope your experience with SEAHF 2020 is a fruitful and long lasting one.

Efforts taken by peer reviewers contributed to improve the quality of papers by providing constructive critical comments; improvements and corrections for the authors are gratefully appreciated. Many thanks to our Session Chairs and for supporting TCT&S2AI&GETS&DMST Event. With their contribution, it was a successful event. Also, we are thankful to all the authors who submitted papers, because of which the conference became a story of success. It was the quality of their presentations and their passion to communicate with the other attendees that really made this conference series a grand success.

Last but not least, we are very grateful for the enormous and continuous support of Slama Best Choice event Organization Company for their support in every step of our journey towards success. Its support was not only a strength but also an inspiration for organizers.

We wish all attendees of SEAHF 2020 an enjoyable scientific gathering. We look forward to seeing all of you next year at the conference.

The editors

Section I: Telecommunications and computing technologies

Innovative and Intelligent Technology-Based Services for Smart Environments – Ben Slama et al (eds)
© 2021 Taylor & Francis Group, London, ISBN 978-1-032-02030-3

Compact and multiband CPW printed monopole antenna based on CLL elements for wireless applications

M. Smari, S. Dakhli & F. Choubani
University of Carthage, SUPCOM, Ariana, Tunisia

J.M. Floc'h
IETR, INSA, Rennes, France

ABSTRACT: The design and performance characteristics of single, dual and tri-band printed CPW monopole antenna covering UMTS (1.93–1.99 GHz), ISM (2.43–2.4835 GHz), WLAN 2.4 GHz, WiMAX 2.5 GHz and S-band 3 GHz frequencies band are reported. These antennas are based on a coaxially fed printed monopoles integrated in a planar configuration with capacitively loaded loops (CLLs) as its near-field resonant parasitic radiators (NFRP). In the proposed technology, the number of operating frequencies is determined by the number of CLLs. These antennas are geometrically simple and have a compact size. The simulated results of return loss, surface currents and radiation patterns are presented and discussed.

1 INTRODUCTION

The interest in and demand for wireless and mobile platforms has grown dramatically. Applications include, for example, biosensors, RFID tags, radars, health applications and crop monitoring and communication systems. Consequently, the need and desire for multifunctional and compact antennas [1–8] has also blossomed, and research in this area has flourished.

In this paper, we propose a multiband printed monopole antenna composed with a monopole antenna associated to Capacitively Loaded Loop (CLL). This study covers the monopole alone then adds CLLs to the basic structure in order to grant multi-frequency behavior. The proposed antenna is suitable for multi-standard wireless communication applications.

2 PROPOSED ANTENNAS DESIGN

2.1 Basic structure: Monopole alone

The design procedure starts with the basic structure (Figure 1). It consists on a monopole antenna of length 32.5 mm and width of 1.2 mm, printed on a $35 \times 35\,\text{mm}^2$.

The resulting antenna was fabricated with an FR4 substrate (dielectric constant $\varepsilon_r = 4.6$, tan $\delta = 0.02$ and thickness h = 0.8 mm) and fed by a CPW transmission line with the input impedance of 50Ω. The monopole is designed to operate at 2.45 GHz. The structure was simulated by using the ANSYS-HFSS high frequency structure simulator.

2.2 Monopole with one CLL element

A Capacitively Loaded Loop (CLL) is placed in the near field region of the printed monopole in order to generate an additional frequency lower than of the driven element. The printed monopole is designed to operate at $F_1 = 2.45$ GHz in the ISM band without any additional matching network.

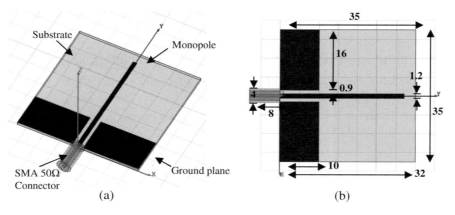

Figure 1.   Antenna design (a) Schema of the structure under HFSS. (b) Geometry and dimensions (in mm).

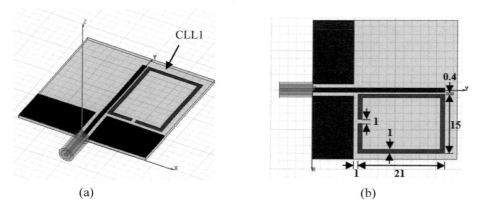

Figure 2.   Proposed antenna design with one CLL element: (a) Schema of the structure under HFSS (b) Geometry and dimensions (in millimeters) of the antenna.

The CLL is designed with an optimized distance between the monopole arm and the ground strip to operate at $F_2 = 0.9$ GHz of GSM band.

2.3   Monopole with two CLL elements

Two Capacitively Loaded Loops (CLLs) are placed in the near field region of the printed monopole in order to generate an additional frequency lower than the driven element. In addition to the structure described in last section 2.2, a second CLL element (CLL2) is inserted at the left side of the monopole and designed to operate at $F_3 = 2$ GHz of UMTS band.

3   SIMULATION RESULTS

3.1   Single-band antenna

3.1.1   Return loss
The simulated return loss shows the resonance frequency of the monopole $F_1 = 2.45$ GHz (Figure 4).

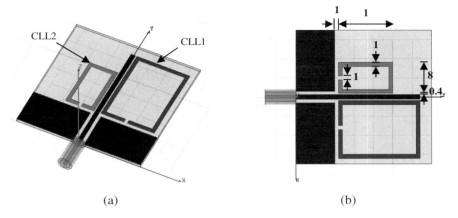

(a)   (b)

Figure 3.   Proposed antenna design with Two CLL elements (a) Schema of the structure under HFSS. (b)
Geometry and dimensions (in mm) of the antenna.

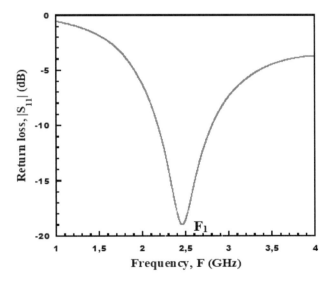

Figure 4.   Simulated return loss $|S_{11}|$ of the single-band antenna.

   The operating frequency band simulated for the single-band antenna is 2.2–2.7 GHz, which can
apply to WLAN 2.4 GHz and WIMAX 2.5 GHz.

3.1.2 *Radiation patterns*
The simulated 3D-radiation pattern of the monopole antenna at the resonance frequency F1 is given
in Figure 5.

3.2 *Dual-band antenna*

3.2.1 *Return loss*
The simulated return loss of the antenna with the insertion of one CLL is presented in Figure 6.

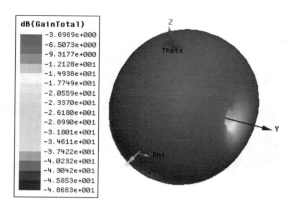

Figure 5.   Simulated 3D-radiation pattern of the proposed antenna at the resonance frequency $F_1 = 2.45$ GHz.

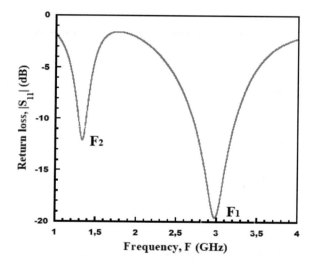

Figure 6.   Simulated return loss $|S_{11}|$ of the dual-band antenna.

We can conclude from Figure 6 that a second resonance frequency $F_2 = 1.33$ GHz is generated in addition to the first resonant frequency $F_1 = 2.45$ GHz of the reference antenna when we insert a CLL. According to presented results, two observations can be made:

– The operating frequency F_1 of the basic antenna is slightly shifted from 2.45 to 2.99 GHz and remains unchanged. This shift of frequency can be explained by the coupling effect between the monopole and CLL element.
– The additional frequency F_2 is shifted to a lower frequency.

In the dual-band mode, the simulated frequency bands are 1.28–1.37 GHz/2.7–3.2 GHz.

3.2.2   Radiation patterns
The simulated 3D-radiation patterns of the structure are given in Figure 7. These patterns correspond to the resonant frequencies F_1 and F_2.

We can notice from Figure 7 that the antenna presents two radiation behaviors. At the frequency F_1 (Figure 7a), the structure is characterized by an axially symmetric radiation pattern around the y-axis. At this frequency, the radiation patterns correspond to the basic printed monopole.

Whereas, and according to Figure 7b, the structure exhibits a relatively directive radiation pattern between x and y-axis at the frequency F_2. We can conclude that the radiation pattern's shape may be a superposition between the electric dipole (monopole) and magnetic dipole (CLL) element.

3.3 *Triple-band antenna*

3.3.1 *Return loss*

The simulated return loss of the antenna with the insertion of two CLLs is presented in Figure 8.

By adding a second CLL to the dual-band antenna seen in section 3.2, we notice that a third resonance frequency F_3 is generated.

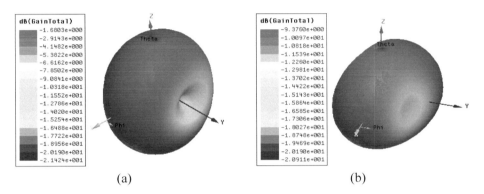

Figure 7. Simulated 3D-radiation pattern of the proposed antenna (a) at $F_1 = 2.99$ GHz; (b) at $F_2 = 1.33$ GHz.

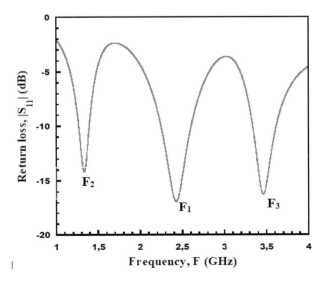

Figure 8. Simulated return loss $|S_{11}|$ of the triple-band antenna.

According to presented results, three observations can be made:

– The operating frequency F_1 of the basic antenna remains almost unchanged (variation of 0.01 – 0.03 GHz).
– The resonance frequency F_2 of the first CLL remains unchanged and equal to 1.33 GHz
– The additional frequency $F_3 = 3.48$ GHz is shifted to a higher frequency upper than the basic monopole antenna

In the triple-band antenna, the simulated frequency bands are 1.2–1.3 GHz/ 2.2–2.6 GHz/ 3.3–3.6 GHz, which can cover the L-band 1GHz/ISM, WLAN 2.4 GHz and WIMAX 2.5 GHz/S-band 3GHz.

3.3.2 Surface currents

In order to analyze the origin of these resonances, we have studied the distribution of the surface currents on the copper elements for proposed structure at the resonance frequencies $F_1 = 2.42$ GHz, $F_2 = 1.33$ GHz and $F_3 = 3.46$ GHz

From Figure 9(b), it is verified that the lowest frequency resonance at $F_2 = 1.33$ GHz corresponds to resonance of the larger (right) CLL1. Similarly, in Figure 9(c), the frequency $F_3 = 3.46$ GHz corresponds to the smaller (left) CLL2 element.

The resonance $F_1 = 2.42$ GHz (Figure 9-a) clearly shows the monopole behaviour. We can notice that when each of the structures is resonant, the others are not.

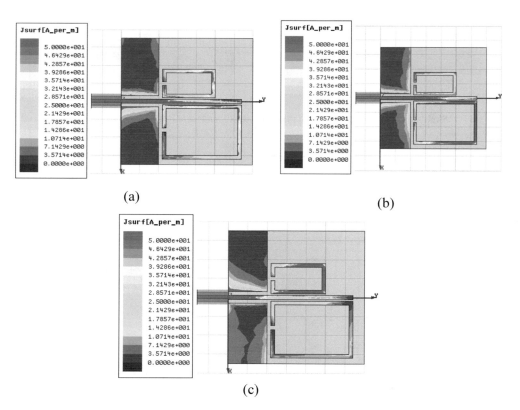

(a)

(b)

(c)

Figure 9. Simulated surface current corresponding to: (a) $F_1 = 2.42$ GHz, (b) $F_2 = 1.33$ GHz, (c) $F_3 = 3.46$ GHz.

3.3.3 *Radiation patterns*

The simulated 3D-radiation patterns of the structure are given in Figure 10. These patterns correspond to the resonant frequencies F_1, F_2 and F_3.

We can notice from Figure 10, that the antenna presents two radiation behaviors. At the frequency F_2 (Figure 10a) the structure is characterized by an axially symmetric radiation pattern around the y-axis. We can conclude that the CLL1 acts as a monopole element at this frequency. Whereas, and according to Figure 10b and Figure 10c, the structure exhibits a relatively directive radiation pattern between x and y-axis.

We can conclude that the radiation patterns shape may be a superposition between the electric dipole (monopole) and magnetic dipole (CLL) elements.

We can notice from the Table 1, that two additional resonance frequencies F_2 and F_3 are generated the first one lower than the frequency $F_1 = 2.45$ GHz of the reference antenna by adding one CLL. The second one is higher than the frequency of the reference after adding the second CLL.

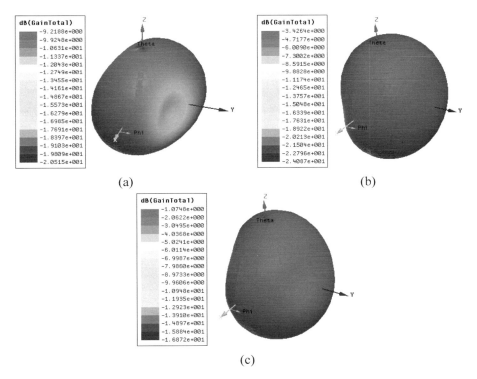

(a)   (b)

(c)

Figure 10. Simulated 3D-radiation pattern of the proposed antenna (a) at $F_2 = 1.33$ GHz; (b) at $F_1 = 2.42$ GHz; (c) at $F_3 = 3.46$ GHz.

Table 1. Resonant frequencies for each antenna type.

	F_1 (GHz)	F_2 (GHz)	F_3 (GHz)
Monopole alone	2.45	–	–
Monopole with 1 CLL	2.99	1.33	–
Monopole with 2 CLL	2.42	1.339	3.46

According to presented results, three observations can be made:

- The operating frequency F_1 of the basic antenna is shifted from 2.45 to 2.99 GHz by adding one CLL. This shift of frequency can be explained by the coupling effect between the monopole and CLL elements.
- The additional frequency F_2 is shifted to lower frequencies while adding one CLL and remains unchanged by adding the second CLL.

4 CONCLUSION

In this work, a family of multifrequency and compact printed antennas is reported. The structure is based on the utilization of capacitively loaded loops (CLLs) placed close to a CPW printed monopole. Three antennas are designed, fabricated and tested. According to the simulated result, the presence of the CLLs generates multi-resonance frequencies covering several standards like UMTS (1.93–1.99 GHz), ISM (2.43–2.4835 GHz), WLAN 2.4 GHz, WiMAX 2.5 GHz and S-band. These properties make the proposed antennas well-suited for emerging multiband wireless applications.

REFERENCES

[1] S. Dakhli, J. M. Floch, H. Rmili, K. Mahdjoubi and H. Zangar. 2012, September 02-05. A novel multifrequency and low-profile metamaterial-inspired monopole antenna, *Mediterranean Microwave Symposium, MMS*. Istanbul:Turkey.

[2] S. Dakhli, J-M. Floc'h, K. Mahdjoubi, H. Rmili and H. Zangar. 2013, April 08-12. Compact and multiband metamaterial-inspired dipole antenna, *European Conference on Antennas and Propagation, EuCAP*. Gothenburg Sweden.

[3] O. Turkmen, G. Turhan-Sayan, and R. W. Ziolkowski. 2014. Single-, dual-, and triple-band Metamaterial-inspired electrically small planar magnetic dipole antennas, *Microw Opt Technol Lett* 56, 83–87.

[4] G. Chaabane, C. Guines, M. Chatras, V. Madrangeas, P. Blondy. 2015. Reconfigurable PIFA Antenna Using RF MEMS Switches, *9th European Conference on Antennas and Propagation (EuCAP)*.

[5] S. Dakhli, H. Rmili, J. M. Floc'h, M. Sheikh, A. Balamesh, K. Mahdjoubi, F. Choubani and R. Ziolkowski. 2016. Printed Multiband Metamaterial-Inspired Antennas with Shaped Radiation Patterns, *Microwave and optical technology letters, MOTL, Vol.* 58, *No.* 6, *pp* 1281–1289.

[6] L. Ge and K. M. Luk. 2016. Band Reconfigurable Unidirectional Antenna: A simple, efficient magneto electric antenna for cognitive radio applications, *IEEE Antennas and Propagation Magazine, vol.* 58, *pp.* 18–27.

[7] M. Kim, L. Qu, H. Shin and H. Kim. 2019. Performance enhancement of dual-band antenna for internet of things applications using closed loops, *Electronics Letters,* Volume: 55, Issue: 25.

[8] Y. Tusharika, S. Sreelekha and Pranamika Balaji. 2019. Dual-Band Complementary Slot Fed Antenna for RFID Applications, *Fifth International Conference on Science Technology Engineering and Mathematics (ICONSTEM)*. Chennai: India.

Innovative and Intelligent Technology-Based Services for Smart Environments – Ben Slama et al (eds)
© 2021 Taylor & Francis Group, London, ISBN 978-1-032-02030-3

Design of a Double Rectangular Spiral Reader Antenna (DRSA) for RFID UHF near-field applications

A. Darghouthi
Higher Institute of Informatics and Multimedia of Gabes (ISIMG), Tunisia

M. Dhaouadi & F. Choubani
Innov'COM Laboratory, Higher School of Communications of Tunis (SUPCOM), Ariana, University of Carthage, Tunisia

ABSTRACT: This paper presents a double spiral near field UHF RFID reader antenna with low Far-Field Gain. The antenna prototype is printed on to a piece of FR4 substrate with an overall size of $111 \times 121 \times 16\,mm^3$ The proposed antenna is composed of a power divider, phase shifter and two lines finished by two symmetric spirals. It generates a strong uniformly distributed magnetic field over a large interrogation. The proposed antenna has low far-field gain and broadband characteristics. The near field operation is achieved by introducing a 90° phase shift between the currents flowing along the opposite side of two branches. An antenna with a spiral of six turns has been fabricated and measured. The experimental results show that the antenna achieves the impedance matching over the frequency range from 800 to 1000 MHz The measured tests on reading range are carried out, and reading distance of far-field half-wave dipole antenna achieves a reader range of 20 cm.

Keywords: Radio Frequency Identification (RFID), Reader, Ultra High Frequency (UHF), Antenna, near field, Opposite Directional Currents (ODCs).

1 INTRODUCTION

UHF RFID systems are being more and more attractive in industrial applications and services. UHF RFID systems may use in two types of applications which are near-field and far-field applications. RFID near-field systems may be used in the management of items (desktop scenes or smart shelves), product tracking, supply chains and bio-sensing applications (Nikitin et al. 2007, Jaakkola & Koivu 2015, Dhaouadi et al. 2013, Jaakkola 2016). The reader antenna and tags may couple by one of two coupling types which are magnetic coupling (inductive) and electrical coupling (capacitive) (Dhaouadi et al. 2015). The magnetic coupling is known when most of the reactive energy is stored in magnetic field (Dhaouadi et al. 2010). The electrical coupling is known when the most of the energy is stored in electric field. The magnetic coupling is the most attractive because it is less sensitive of metals and liquids (Amendola et al. 2015, Dhaouadi et al. 2014). The performance of RFID systems is based extremely on the reader antenna. To design an UHF RFID near field magnetic coupling reader antenna and control field distribution is becoming the greatest challenge nowadays. It requires the antenna distribution field concentrated in near field and decreases gradually in far field. Y. Yao (Yao et al. 2017) designs a meander line UHF RFID reader antenna which exhibits uniform electric and magnetic fields with low cost and an interrogation zone controlled by adjusting the number of microstrip units or the output power, which offers a maximum reading range of 30 cm. He designed also a multi-polarized RFID reader antenna for UHF near-field applications presented in Yao et al. 2017. This antenna generates a strong and uniform electric field and can detect linear polarization tags which are orientated arbitrarily and achieve a maximum reading range of 35 cm.

A dual-log-periodic array antenna (LPDA) is designed in (Li et al. 2018) for RFID UHF near field applications. The concept of ODCs is applied in this antenna to enhance the magnetic field in the near field, which offers a maximum reading range of 19 cm. An embeddable reader antenna is simulated and prototyped for billing kiosk application in a retail environment (Parthiban 2019). The fabricated antenna is measured in terms of different RF (Radio Frequency) and RFID performance metrics and shows that it meets all the requirements sought by application.

In this research, a dual rectangular spirals reader antenna with low far field gain is proposed for RFID UHF near-field applications. The proposed antenna is characterized by the strong and uniform magnetic field distribution in near field, which attenuated rapidly in the far-field. In addition, the measured impedance matching -10 dB achieves three band covers the UHF band, which are [864–880], [925–938] and [970–988] in an operating frequency equal to 867 MHz. This antenna is simple in fabrication and offers a maximum reading range of 20 cm using the half-wave dipole antenna. The performance of the proposed antenna is evaluated by the measurement results.

2 ANTENNA CONFIGURATION

The configuration of the proposed rectangular spirals reader antenna is shown in Figure 1.

The antenna is comprised of a phase shifter with two stubs, power dividers and double symmetric microstrip lines finished by double symmetric spirals, as shown in Figure 1a. The antenna is fabricated on an FR4 substrate which have a permittivity equal to 4.4, thickness equal to 1.6 mm and dimensions (L,W). The prototyped antenna is based on the principle of ODCs (Opposite Directional Currents) as explained in Yao et al. 2017. The spiral has a length approximately equal to $4\lambda_g$, where λ_g is the operating frequency. The phase shifter having a length of $\lambda_g/3$ after optimization. As shown in Figure 1b, the proposed antenna is connected to an SMA connector which has an impedance equal to 50 Ω. Table 1 presents the optimized dimensions of the antenna. The equation 1 presents the magnetic field in function of turns number of spiral N:

$$H = \frac{\mu_0 \times N \times I}{L} \tag{1}$$

with μ_0: permeability, N is the number of spiral turns, I is the current and L is spiral length.

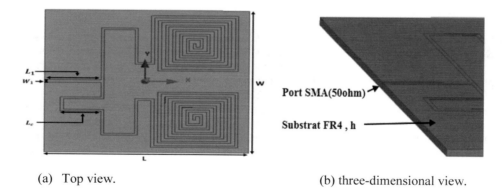

(a) Top view. (b) three-dimensional view.

Figure 1. Configuration of the proposed antenna.

Table 1. Optimized dimensions of the proposed antenna (mm).

W	L	W_1	L_1	L_c
121	111	2	21.8	22

As the number of turns increases, the magnetic field increases. By using spirals we can control the magnetic field by changing the number of turns. Which show the use of the spirals in our antenna. The principle of the proposed antenna is explained in the following sections.

3 SIMULATION RESULTS

The proposed antenna was optimized and designed using HFSS (High Frequency Structure Simulator), which is based on the finite elements method and is a full wavelength numeric electromagnetic simulation tool.

3.1 *The impact of turns number of spirals*

We can notice that the optimal case is for 6 turns. The strength of magnetic field along the x axis has a maximum in −2.18 dB, as shown in Figure 2a. The strength of magnetic field along y axis has a maximum in 10.21 dB, as shown in Figure 2b. Which indicates that the antenna is feasible in near field applications. The simulated strength of magnetic field in the center of antenna with

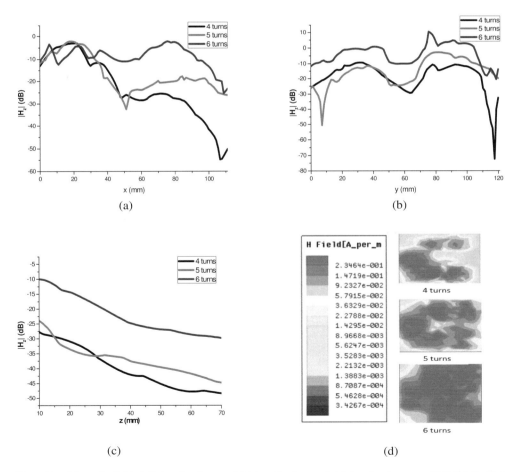

Figure 2. Magnetic field: (a)intensity along x, (b) intensity along y, (c) intensity along z and (d) distribution.

respect to the distance along Z-axis, is strong in near field with a maximum value of -10 dB for 10 mm and weakened greatly in far field, as shown in Figure 2c.

3.2 *The impact of vertical offset z*

The magnetic field distribution and intensity for three value of vertical offset z (z = 10 mm, 30 mm and 50 mm), presented in Figure 3, for an operating frequency equal to 867 MHz and indicates that the proposed antenna produces a uniform magnetic field in a controlled interrogation zone and drops quickly when the vertical distance increase.

3.3 *Current distribution*

Figure 4 depicts the current distributions in the antenna structure at different phases ($0°$, $90°$, $180°$, $270°$). It can be seen that the current direction changed with the phase. So the phase shifter causes a phase difference of $90°$ between the lower and upper branches. Moreover, the $90°$ phase difference

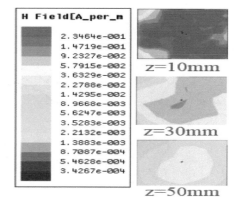

Figure 3.   Magnetic field for z=10 mm, 30 mm, and 50 mm.

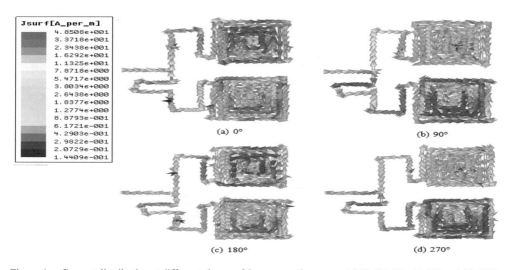

Figure 4.   Current distribution at different phases of the proposed antenna: (a) $0°$, (b) $45°$, (c) $90°$ and (d) $270°$.

14

between the lower branch and the upper branch has generated a strong magnetic field intensity on this antenna.

4 MEASUREMENTS RESULTS

The antenna is fabricated to validate the simulation results and to evaluate his performance. The overall size of the fabricated antenna is 111 mm × 121 mm × 1.6 mm and copper layer on the board with a thickness equal to 0.035 mm as shown in Figure 5.

We used the Agilent 8714ES vector network analyzer to measure the impedance matching of the antenna. Simulated and measured return loss are shown in Figure 6.

The frequency bands achieved for S11<−10 dB range from [864 MHz–880 MHz] at a resonant frequency of 871 MHz, [925 MHz–938 MHz] at a resonant frequency of 931 MHz and [970 MHz–988 MHz] at a resonant frequency of 976 MHz, which covers the UHF band. There is an agreement between the simulated and measured impedance matching. The shift between simulated and measured results may be caused by fabrication errors and inconsistence of FR4 substrate.

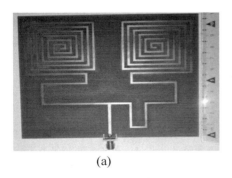
(a)

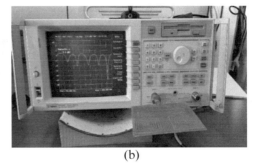
(b)

Figure 5.   Proposed loop antenna for UHF near-field RFID application; (a) photo of the antenna prototype, (b) S11 measurement setup.

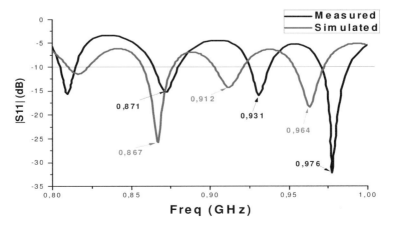

Figure 6.   S parameter of the proposed antenna.

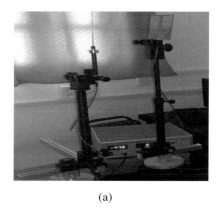

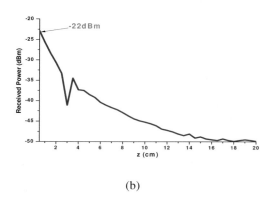

(a)   (b)

Figure 7.   Received power of the proposed antenna: (a) measurement banc, (b) Power Received (dBm).

To further evaluate the performance of the proposed antenna, we take a step length of 0.5 cm between the proposed antenna and the half-wave dipole antenna to measure and evaluate the maximum reading range as a function of z. As can be seen in Figure 7, the received power achieves this threshold which is equal to −50 dBm within 20 cm using a half-wave dipole antenna. As the distance between the proposed antenna and the receiver antenna increases, as the received power decreases. it is clear that reading scope is reduced if distance is increased. The received power decreases irregularly because of the presence of interference in the detection environment.

5   CONCLUSION

In this paper, a novel double rectangular spirals reader antenna is presented for RFID near-field applications. The proposed antenna has many characteristics such as a simple structure, a low cost and easy integration with other devices. The proposed loop antenna has demonstrated the capability of producing strong uniform magnetic fields in the near-field. Both simulated and measured results show good capabilities for near-field applications. Such a novel rectangular spiral antenna is suitable for near-field RFID operations.

REFERENCES

Amendola, S. Milici, S., & Marrocco, G. (2015), Performance of Epidermal RFID Dual-loop Tag and On-Skin Retuning, *IEEE Transactions on Antennas and Propagation*, 63, 3672–3680. DOI: 10.1109/TAP.2015. 2441211

Dhaouadi, M. Mabrouk, M. Vuong, T. P. de Souza, A. C., & Ghazel, A. (2014), UHF tag antenna for near-field and far-field RFID applications, *15th annual IEEE Wireless and Microwave Technology Conference (WAMICON 2014)*, Tampa, FL, 1–4. DOI: 10.1109/WAMICON.2014.6857794

Dhaouadi, M. Mabrouk, M. Vuong, T. P. de Souza, A. C., & Ghazel, A. (2015), A capacitively-loaded loop antenna for UHF near-field RFID reader applications, *Radio and Wireless Symposium (RWS)*, San Diego, CA, 193–195. DOI: 10.1109/RWS.2015.7129755

Dhaouadi, M., Mabrouk, M. Vuong, T. P. Hamzaoui, D., & Ghazel, A. (2013), Chip impedance matching For UHF-band RFID TAG, *7th European Conference on Antennas and Propagation (EuCAP)*, Gothenburg, 3056–3059.

Dhaouadi, M. Mabrouk, M., & A. Ghazel, (2010), Magnetic antenna for near-field UHF RFID tag, *18th International Conference on Microwaves, Radar and Wireless Communications*, Vilnius, Lithuania, 1–2.

Jaakkola, K. (2016), Small On-Metal UHF RFID Transponder With Long Read Range, *IEEE Transactions on Antennas and Propagation,* 64, 4859–4867. DOI: 10.1109/TAP.2016.2607752

Jaakkola, K., & Koivu, P. (2015), Low-Cost and Low-Profile Near Field UHF RFID Transponder for Tagging Batteries and Other Metal Objects, *IEEE Transactions on Antennas and Propagation,* 63, 692–702. DOI: 10.1109/TAP.2014.2378260

Li, H. Chen, Y., & Yang, S. (2018), A Novel Printed Dual-Log-Periodic Array Antenna for UHF Near-Field RFID, *IEEE Transactions on Antennas and Propagation,* 66, 7418–7423. DOI: 10.1109/TAP.2018.28 74202

Nikitin, P. V., Rao, K. V. S., & Lazar, S. (2007), An Overview of Near Field UHF RFID, in *IEEE International Conference on RFID,* 167-174. DOI: 10.1109/RFID.2007.346165

Parthiban, P. (2019), Fixed UHF RFID Reader Antenna Design for Practical Applications: A Guide for Antenna Engineers With Examples, *IEEE Journal of Radio Frequency Identification,* 3, 191–204. DOI: 10.1109/JRFID.2019.2920110

Yao, Y. Cui, C. Yu, J., & Chen, X. (2017), A Meander Line UHF RFID Reader Antenna for Near-Field Applications, *IEEE Transactions on Antennas and Propagation,* 65, 82–91. DOI: 10.1109/TAP.2016.2631084

Yao, Y. Liang, Y. Yu, J., & Chen, X. (2017), Design of A Multipolarized RFID Reader Antenna for UHF Near-field Applications, *IEEE Transactions on Antennas and Propagation,* 65, 3344–3351. DOI: 10.1109/TAP.2017.2700873

Innovative and Intelligent Technology-Based Services for Smart Environments – Ben Slama et al (eds)
© 2021 Taylor & Francis Group, London, ISBN 978-1-032-02030-3

A compact MIMO antenna with high isolation for portable UWB applications

M. Dhaouadi & F. Choubani
Innov'COM Laboratory, Higher School of Communications of Tunis (SUPCOM), Ariana,
University of Carthage, Tunisia

ABSTRACT: In this work, a compact ultrawideband (UWB) multiple-input–multiple-output (MIMO) antenna with a high isolation for portable UWB applications is presented. The proposed antenna consists of two monopole antenna element symmetrical squares, with a base plane that makes up the DGS structure to improve the insulation between two antenna elements. The prototype is simulated, fabricated, and measured. A good agreement between the measured and simulated results shows that an ultrawide bandwidth and a high isolation are successfully achieved. The mutual coupling measured for $S_{21} < -20$ dB is from 3 to 10 GHz and for $S_{21} < -40$ dB is 7.4 to 10 GHz. These performances indicate that the proposed MIMO antenna is a competitive candidate for portable UWB applications.

Keywords: Small antenna, Structure DGS, High isolation, Envelope correlation coefficient.

1 INTRODUCTION

In recent years, an ultrawideband (UWB) communication system has been studied to meet the demand for high data rates, low costs and low power consumption. Since the Federal Communications Commission (FCC) authorized the unlicensed 3.1–10.6 GHz band for UWB communications, UWB communications has become a hot topic in the wireless communications arena [1].

The challenge of designing a workable UWB antenna includes broad impedance matching, radiation stability, a low profile, compact size and low cost [2]. In addition, UWB systems also suffer from multi-channel fading like other wireless systems. To solve this problem, the multiple input and multiple output technique (MIMO) is introduced in UWB systems to provide multiplex gain and diversity gain, further improving the capacity and quality of the link [5].

Two major challenges are encountered in the design process of MIMO antennas for UWB systems. One is to minimize antenna elements for MIMO systems. The other is to improve the isolation between the antenna elements. Multi-input multi-output (MIMO) can improve the capacity of the channel and has been used to solve the problem of improving the capacity of the channel and increasing the transmission rate [4–5].

To enhance the performance of the UWB communication system, the UWB technique and the MIMO technique have been combined to create an amazing technique named the UWB-MIMO communication system [4–5]. UWB-MIMO antennas have been becoming a hot topic in recent years. Moreover, several UWB-MIMO antennas have been presented to provide desired communication requirements [5]. However, the coupling between the MIMO antenna elements is always high because of the compact antenna design and the space limitation of the modern portable communication terminals. Many UWB-MIMO antennas with high isolation have been exploited to reduce the coupling between the used antenna elements. Unfortunately, some of the UWB antennas are complex in structure and others are large in size. In [6], a dual band notch was designed for a UWB-MIMO antenna using parasitic strips and slots on the radiator. To the best of the authors' knowledge, it is the smallest one among all the UWB-MIMO antennas with and without notches

DOI 10.1201/9781003181545-3

found in literature. In [7] and [8], slot antennas were designed for UWB MIMO applications with a strip to ensure high isolation. The slots were etched on the feeding structure to create a band notch. However, this kind of structure had a relatively large size.

2 ANTENNA DESIGN AND ANALYSIS

2.1 *Antenna design*

The geometry of the proposed UWB MIMO antenna, with a small size of 12×20 mm^2, is shown in figure 1. It is printed on a FR4 substrate with a relative permittivity of 4.4 and a thickness of h = 1.6 mm. The antenna with a T shape on the ground plane proposed in [5] is used as a reference, and the dimensions of the antenna are optimized to achieve a smaller size. The proposed UWB-MIMO antenna is composed of two patch antennas with L and H-shaped slots placed symmetrically, fed by a line with 50 microrubans to achieve good insulation between the two antenna elements. These radiant elements are placed on the upper layer and on the lower layer we added onto the ground plane the DGS structure composed of rectangular faults to give good insulation.

Improved bandwidth for UWB applications is achieved by adding rectangular bands on the mass plane to improve isolation between low band antenna elements and the DGS structure is used in the form of a "T." To obtain the required numerical analysis and the correct geometry parameters, computer simulation using electromagnetism (EM) and the HFSS simulation tool is performed.

3 RESULTS AND DISCUSSION

3.1 *S-parameters*

A prototype of the MIMO antenna described in Section 2 is fabricated and measured. The prototype is shown in figure 1(b). The dimensions for the final design are listed in table 1. The bandwidth performance of this proposed antenna is measured by the network analyzer (VNA). Figures 2(a) and (b) give the simulated and measured S-parameters of the proposed antenna.

As shown in figure 2(a), the antenna is adapted into two bandwidths, the first band impedance bandwidth measured for $S_{11} < -10$ dB is 3.5 to 6.9 GHz and the second measurement band is 8.7 to

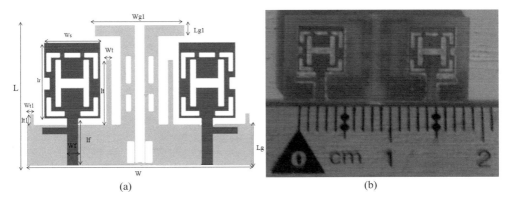

Figure 1. Configuration of the proposed antenna: (a) geometry of the proposed antenna, (b) photograph of prototyped antenna.

Table 1. Dimensions of the proposed antenna (mm).

Parameters	**W**	**L**	**L$_g$**	**l$_t$**	**w$_t$**	**w$_{t1}$**	**w$_{g1}$**	**L$_{g1}$**	**l$_r$**	**w$_s$**
Unit (mm)	20	12	3.5	5.5	0.5	0.4	8	1	6.5	5

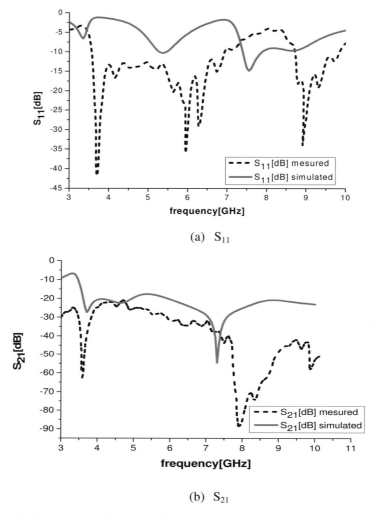

(a) S_{11}

(b) S_{21}

Figure 2.   Simulated and measured (a) S_{11} and (b) S_{21}.

9.8 GHz so this antenna meets the impedance matching requirement for the entire UWB specified by the FCC.

The results of simulated and measured mutual coupling between the two input ports is shown in figure 2(b). This gives a measured mutual coupling $S_{21} < -20$ dB over the entire UWB from 3 to 10 GHz and $S_{21} < -40$ dB is 7.4 to 10 GHz; this band indicates that the insulation between the two antennas is very high. This antenna is then suitable for the MIMO application across the entire UWB band.

There is a difference between the measured and simulated S_{11} and S_{21} due to the power cable used in the measurement.

3.2   Radiation performance

Figure 3 presents the simulated 3-D radiation diagram of the antenna with the different color scales at low frequency at the frequency of 3 GHz and 10 GHz. The maximum gain of this antenna is present at the frequency of 10 GHz equal to 10 dB.

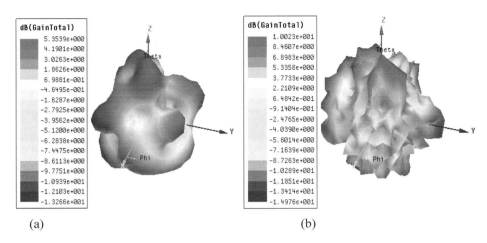

Figure 3.   3-D radiation patterns (a) at 3 GHz, (b) at 10 GHz.

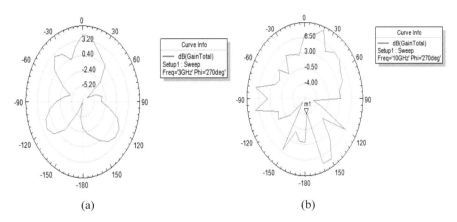

Figure 4.   2-D radiation patterns (a) at 3 GHz, (b) at 10 GHz.

The simulated and measured 2-D radiation patterns at 3 GHz and 5 GHz have good agreements as indicated in Figure 4.

3.3 *Diversity performance*

The antenna used for MIMO application, the two-port envelope correlation coefficient (ECC) is an important parameter. Recall that for a loss-less MIMO antenna, the ECC can be calculated using the method proposed in [9].

$$\rho_e = \frac{\left| S_{11}^* S_{12} + S_{21}^* S_{22} \right|^2}{\left(1 - |S_{11}|^2 - |S_{21}|^2 \right)\left(1 - |S_{22}|^2 - |S_{12}|^2 \right)} \tag{1}$$

The simulated and measured ECC curves are plotted in figure 5. It shows that both the simulated and measured ECC are below 0.24 in 3–10 GHz, which is low enough to ensure good diversity performance for the presented MIMO antenna.

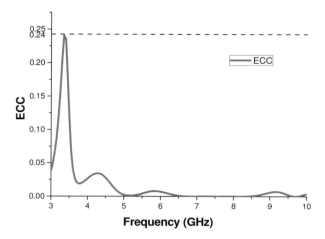

Figure 5.   Simulated envelope correlation coefficients.

4   CONCLUSION

In this article, a UWB MIMO antenna with a very compact a size of 12×20 mm^2 was produced and measured. High insulation between two ports has been achieved by etching H and L slots in the radiator. The measured results show that the proposed antenna reaches an impedance bandwidth greater than 3.5–9.8 GHz and a weak mutual coupling of less than -40 dB on the 7.4 GHz band at 10 GHz and higher bands respectively with an envelope correlation coefficient of less than 0.24. The measurements show that the proposed MIMO antenna is a good candidate for portable UWB applications.

REFERENCES

[1] Federal Communications Commission, 2002, Federal Communications Commission revision of Part 15 of the Commission's rules regarding ultra-wideband transmission system from 3.1 to 10.6 GHz, *ET-Docket*, Washington, pp. 98–153.

[2] F. Zhu et al. 2013, Multiple band-notched UWB antenna with band-rejected elements integrated in the feed line *IEEE Transactions on Antennas and Propagation* vol. 61, no. 8, pp. 3952–3960.

[3] L. Li, S. W. Cheung, and T. I. Yuk, 2013, Compact MIMO antenna for portable devices in UWB Applications, *IEEE Transactions on Antennas and Propagation*, vol. 61, no. 8, pp. 4257–4264.

[4] T. Jiang, T. Jiao, Y. Li, 2016, Array mutual coupling reduction using Loading E-shaped electromagnetic band gap structures,*International Journal of Antennas and Propagation*, vol.2016, pp.1–9.

[5] Y. Li, W. Li, W. Yu, 2013, A multi-band/UWB MIMO/diversity antenna with an enhance isolation using radial stub loaded resonator, *Applied Computational Electromagnetic Society Journal*, vol.28, no.1, pp. 8–20.

[6] J. F. Li, Q. X. Chu, Z. H. Li, and X. X. Xia, 2013, Compact dual band-notched UWB MIMO antenna with high isolation, *IEEE Transactions on Antennas and Propagation.*, vol. 61, no. 9, pp. 475–4766

[7] P. Gao, 2014, Compact printed UWB diversity slot antenna with 5.5-GHz band-notched characteristics, *IEEE Antennas* and *Wireless Propagation Letters*, vol. 13, pp. 376–379.

[8] B. P. Chacko, G. Augustin, and T. A. Denidni, 2013, Uniplanar polarization diversity antenna for wideband systems,*IET Microwaves Antennas & Propagation*, vol. 7, no. 10, pp. 85–857.

[9] S. Blanch, J. Romeu, and I. Corbella, 2003, Exact representation of antenna system diversity performance from input parameter description,*Electronics Letters*, vol. 39, no. 9, pp. 70–707.

Innovative and Intelligent Technology-Based Services for Smart Environments – Ben Slama et al (eds)
© 2021 Taylor & Francis Group, London, ISBN 978-1-032-02030-3

When deep learning meets web QoE

Nawres Abdelwahed
Cosim Lab, Supcom, University of Carthage, Tunisia

Asma Ben Letaifa
Mediatron Lab, Supcom, University of Carthage, Tunisia

Sadok El Asmi
Cosim Lab, Supcom, University of Carthage, Tunisia

ABSTRACT: In this contribution, we selected three parameters of web QoE (TTC, h_score and vote) from a large-scale dataset. These parameters are selected since they are the most related to user engagement. The combination of the three web metrics gives as the Mean Opinion Score (MOS) which illustrates the user experience. Then we visualized the selected parameters as well as the MOS. At first, we manually fixed the parameters weights (α, β and γ). Then, we tried to predict them using Artificial Neural Network (ANN) training algorithm but we had a low accuracy.

Keywords: ANN, user engagement, web QoE, MOS.

1   INTRODUCTION

The number of web pages and sites is constantly increasing, on the one hand. On the other hand, the number of web browser users is also increasing as mentionned in Internet World Stats (2020). Faced with this duo which is increasing in a fast way and the high demands of users, network operators as well as service providers are faced with a difficult situation; they must overwork all the difficulties in order to guarantee the quality of their services and satisfy users. To have a good user experience, operators need to know how to predict the feelings of users. Now, predicting the feelings of a human being is a paradoxical sentence in its appearance, but in artificial intelligence anything is possible like mentioned in Russell et al. (2016). Machine Learning is the invention that saves the lives of operators. Using the mechanisms of ML, the prediction of the feelings of users is no longer a wish but it is a concrete reality.

Deep learning as used in the following articles Hinton et al. (2006), Schmidhuber (2015) and Yu & Deng (2011) nowadays is very popular thanks to its brilliant achievements. We notice that deep learning (DL), in past years, succeeded in various application domains especially in image as shown in Krizhevsky et al. (2012) and speech recognition as mentioned in Hinton et al. (2012), Seide et al. (2011) and Dahl et al. (2012). Deep learning is proved able to solve complex tasks while delivering convincing results as explained in Covington et al. (2016), that is why both industry and academia are applying it increasingly to a wider range of applications. Lately, deep learning ameliorates the performance of recommenders in a spectacular way and brings rectifications to recommendation architectures. Recent ameliorations in recommender systems based on deep learning are overcoming conventional models obstacles and obtaining a high quality of recommendation. Deep learning can catch non-linear and non-trivial user-element relationships, and codifies complex abstractions as data representations in the upper layers. In addition, it captures complex relationships in the data itself from accessible and abundant data sources such as textual, visual and contextual information.

In this contribution, we developed at first time a web application, so we can collect data with, and to have as a result our own dataset. However, applying deep learning algorithms on data needs a large-scale dataset which we can not collect in few time, that is why we decided to use an open source dataset of Gao et al. (2017) in this work and our dataset in a future work. From the open source dataset we selected the web metrics that reflects the most the user engagement. Then we combined them with their coefficients to predict the Mean Opinion Score (MOS). To finish, we applied ANN on our modified dataset to predict our parameters coefficients, but low accuracy led us to create our own large-scale dataset so we can apply deep learning algorithms on.

The remainder of this paper is organized as follows. Section 2 defines our approach. In section 3 we present our results. Section 4 is interested in deep learning and its application. And the section 5 concludes our work and open new perspectives.

2 PROPOSED APPROACH

In this section we are going to detail our approach step by step; from the development of our web application to the used algorithm to determine the Mean Opinion Score (MOS).

2.1 *Application*

Figure 1 and Figure 2 show our web application different interfaces. Figure 1 is our home interface, the first showed page when opening the application. Figure 2 is the information related to a participant interface. Figure 3 is the participant evaluation interface. Besides, we have a web pages database with 50 pages selected from the most visited sites according to Alexa's top sites. Figure 3 illustrates some of the selected web pages to be evaluated. Each user is called to note each web page with stars, each page has five stars, i.e., 1, 2, 3, 4, and 5. The five stars 1, 2, 3, 4, and 5 are named as follows: "very low quality", "low quality", "medium quality", "good quality" and "excellent quality".

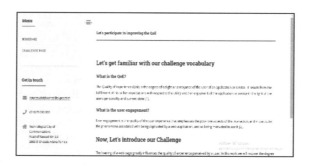

Figure 1.   Home page.

Figure 2.   Challenge page.

Figure 3. Some selected web pages to be noted.

2.2 *Dataset*

One of the challenges of data-based QoE estimation is the lack of ample training data, notably when executing deep learning algorithms. That's why, we need a large-scale QoE dataset to facilitate research on QoE evaluation.

For this reason and for the lack of time to harvest a large-scale QoE dataset we used the dataset of Gao et al. (2017) which is an open source dataset. The used dataset is harvested from a web application whose concept is similar to ours. The used application in Gao et al. (2017) uses videos in their web pages to evaluate the QoE of a web page.

Videos are displayed in pairs, side by side, so the user can decide which of them is the best and note according to that. These videos show how the browser displays above-the-fold content. The used dataset collected its lines from an heterogeneous population (friends, colleags, social media channels), however the community of web performance is targeted and not normal people so the results would be correct. In addition, to ensure a good quality of data, researchers set a series of validation mechanisms. The used dataset is composed of five files (in CSV format). Each one of them contains metrics extracted from WebPageTest on URLs or collected data from participants. The files are csv1.csv (115 rows and 32 columns), csv2.csv (160 rows and 3 columns), csv3.csv (44352 rows and 5 columns), csv4.csv (2772 rows and 2 columns) and finally csv5_SIandPSI.csv (160 rows and 5 columns).

2.3 *Our contribution*

In this work, we are interested in enhancing user engagement metrics to understand the final user in a better way and to collect more real results. That is why, we selected the most relevant QoE metrics, present in the dataset Gao et al. (2017), with a strong correlation to user engagement. As a result, we have three parameters which are: TTC (Time To Click), h_score and vote.

In the first place, we visualize the selected parameters. Then, we try to apply Artificial Neural Network (ANN) algorithm to predict the MOS resulting of the combination of these parameters. After that, we try several different combinations of coefficients assigned to parameters. To finish, we choose the best combination that gives as the more realistic result. Formula 1 gives us a better explanation of the parameters and their coefficients.

$$\alpha A + \beta B + \gamma C = MOS \tag{1}$$

A: TTC
B: h_score
C: vote
α, β and γ are respectively the coefficients of A, B and C

3 RESULTS

In this section we present our results; the visualization of our selected parameters (Figures 4–7), the visualization of the resulting MOS with manually fixed coefficients (Figures 8 and 9) and the visualization of the application of ANN on our new dataset (Figure 10).

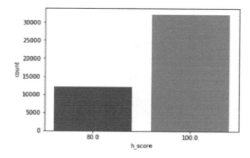

Figure 4. Visualization of vote values.

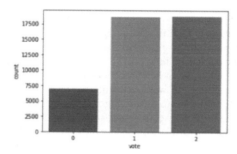

Figure 5. Visualization of h_score values.

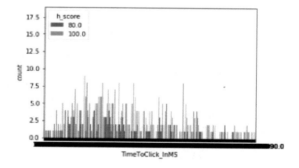

Figure 6. TTC as a function of h_score.

Our aim is to predict the MOS, to do so we calculate the formula (1) defined in section 2. We try different values for our coefficients α, β and γ. As a result, we used this formula: 0.6*TTC+0.5*H_score+0.1*vote=MOS. The results of this formula are presented in the column results. Then, the results are classified in 5 classes from 1 (worst experience) to 5 (best experience). The obtained results are illustrated in Figure 8. Then, Figure 9 shows the obtained MOS classified into 5 levels.

26

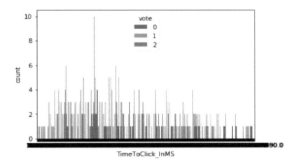

Figure 7.   TTC as a function of vote.

lt_1	PSI_plt_2	PSI_ttc_majority_1	...	uid_left	uid_right	sessionID	TimeToClick_InMS	userAgent	vote	h_score	majority_pick	resultat	MOS
838	3742	1838	...	160605_JY_2PB	160605_X9_2DH	58aPyrqJm6zGYK6eJ	5643.0	Mozilla/5.0 (iPhone; CPU iPhone OS 9_3_2 like ...	1	100.0	1	1138.7	3
742	2446	3742	...	160605_X9_2DH	160605_1W_2ZH	58aPyrqJm6zGYK6eJ	4561.0	Mozilla/5.0 (iPhone; CPU iPhone OS 9_3_2 like ...	2	100.0	2	922.4	2
350	1580	2350	...	160605_TF_2MS	160605_HA_2PK	58aPyrqJm6zGYK6eJ	5484.0	Mozilla/5.0 (iPhone; CPU iPhone OS 9_3_2 like ...	0	100.0	2	1106.8	3
200	1774	6230	...	160605_QJ_2BZ	160605_6Q_311	58aPyrqJm6zGYK6eJ	5119.0	Mozilla/5.0 (iPhone; CPU iPhone OS 9_3_2 like ...	2	100.0	2	1034.0	3

Figure 8.   Obtained MOS results.

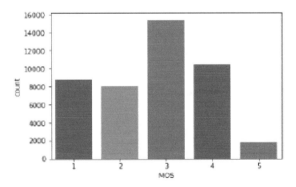

Figure 9.   Visualization of MOS values.

4   DEEP LEARNING

4.1   *Modeling with Artificial Neural Networks*

In Figure 10, we have our proposed ANN model; which is McCulloch & Pitts (1943) model adjusted for our problem: Our input parameters A, B and C each with their respective weights α, β and γ, then they are summed up in the unit frame, pursued by the nonlinear activation function ϕ; it determines the unit when to shoot.

27

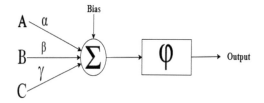

Figure 10.   Proposed Artificial Neural Network model.

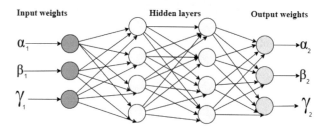

Figure 11.   ANN structure of our proposed model.

Figure 11 presents a network of interconnected neurons from an input layer to an output layer passing by hidden layers like described in Rosenblatt (1958). This configuration is unique; it defines feed-forward networks and differentiates them from other network types such as recurrent networks or self-organizing maps. The MultiLayer Perceptron (MLP) is a multiple hidden layers network. Feedforward networks stand for connections between input and output. Training the ANN algorithms signifies defining different layers weights based on experimental data. In the case of supervised learning, the desired results are already known.

4.2   *ANN application*

We applied ANN on the modified dataset to determine α, β and γ automatically, and as a result we obtained a very low accuracy which is not satisfying. That is why our goal is to ameliorate the accuracy by setting up of a new large-scale dataset. The new dataset will integrate user behavior/user engagement parameters and will help us run the deep learning algorithms and give a better accuracy. We are actually setting up a large dataset.

5   CONCLUSION

In this paper we selected the most expressive QoE parameters and the closest to user engagement to predict the most real MOS possible. The weights associated to our parameters are fixed manually in the first place. Then, we tried to automatically fixe α, β and γ using deep learning (ANN in our case), but we obtained a low accuracy. That is why, we opened new perspectives; we will set our own large-scale dataset with different types of metrics: QoS metrics, web QoE metrics and user engagement metrics.

REFERENCES

Covington, P. et al. (2016). Deep neural networks for youtube recommendations. *In: 10th Proceedings of ACM Conference on Recommender Systems (RecSys)*: 191–198.

Dahl, G. E. et al. (2012). Context-Dependent Pre-Trained Deep Neural Networks for Large-Vocabulary Speech Recognition. *IEEE Transactions on Audio, Speech Recognition and Language Processing(20)*: 30–42.

Gao, Q. et al. (2017). Perceived performance of top retail webpages in the wild. *Insights from large-scale crowdsourcing of abovethe-fold qoe. In: Proceedings of SIGCOMM Internet-QoE workshop.*

Hinton, G. E. et al. (2006). A fast learning algorithm for deep belief nets. *Neural computation(17)*: 1527–1554.

Hinton, G. et al. (2012). Deep neural networks for acoustic modeling in speech recognition: The shared views of four research groups. *IEEE Signal Processing Magazine(29)*: 82–97.

Internet World Stats, https://www.internetworldstats.com/stats.htm, last accessed 2020/03/03.

Krizhevsky, A. et al. (2012). Imagenet classification with deep convolutional neural networks. *In: 25th Proceedings of Advances in neural information processing systems (NIPS)*: 1097–1105.

McCulloch, W. S. & Pitts, W. (1943). A logical calculus of the ideas immanent in nervous activity. *The bulletin of mathematical biophysics(5)*: 115–133.

Rosenblatt, F. (1958). The perceptron: A probabilistic model for information storage and organization in the brain. *Psychological Review(65)*: 386–408.

Russell, S. J. et al. (2016). Artificial Intelligence: a Modern Approach. *2nd edn. Englewood Cliffs, Prentice Hall.*

Schmidhuber, J. (2015). Deep learning in neural networks: An overview. *Neural Networks(61)*: 85–117.

Seide, F. et al. (2011). Conversational speech transcription using context-dependent deep neural networks. *In: 12th Annual Conference of the International Speech Communication Association Florence, Italy*: 437–440.

Yu, D. & Deng, L (2011). Deep learning and its applications to signal and information processing [exploratory dsp]. *IEEE Signal Processing Magazine(28)*: 145–154.

Innovative and Intelligent Technology-Based Services for Smart Environments – Ben Slama et al (eds)
© 2021 Taylor & Francis Group, London, ISBN 978-1-032-02030-3

Using IoT in e-commerce for buying according to your needs

M. Shili & K. Sethom
Innov'COM at Sup'Com Laboratory, University of Carthage, Tunisia

ABSTRACT: Nowadays, IoT (Internet of Things) devices are used everywhere in many domains including e-commerce, eHealth, smart cities, agriculture, etc. The application of RFID technology in e-commerce helps to facilitate communication between suppliers and customers. The key of modern e-commerce is the automatic "id-ing" of products, shopping carts and other staff in the store. The application of an e-commerce system based on RFID technology can fundamentally solve the problem of inventory sales and manage the purchasing process. In this work, we present the application of RFID technology and the internet of things based on an e-commerce system. The RFID in e-commerce is the latest technology used in commerce for managing automated products and also detecting each attempt at stealing or loss of product.

1   INTRODUCTION AND MOTIVATION

The internet of things or IoT as it is known, means taking all the things in the world and connecting them to the internet. It is considered the revolutionary technology of the 21st century in the digital World due to its radiance in all areas and specialties such as the healthcare domain, agriculture, transport and e-commerce in Hanna Mohammad Said & Salem (2019).

We can clearly see how the IoT, this phenomenal technology, makes our daily lives easier and much more comfortable today by guiding the consumer to select the best product with the most appropriate price without any intervention of the seller in a very smart way through the IoT and RFID technology (Atlam et al. 2018). In fact, it is particularly characterized by its huge potential which makes all these smart devices connected and interact with each other for the purpose of sharing and exchanging data with other devices and systems over the internet in a short time. In addition, the internet of things has been significantly influencing a variety of sectors. The internet of things creates a suitable the atmosphere for creativity and productivity in the e-commerce industry (Yunusa & Anas 2018).

2   IOT ARCHITECTURE

Figure 1 illustrates an example of simple architecture in an IOT scenario which can be divided in three layers: the perception layer, the network layer and the application layer (through pc or mobile phones). The IoT architecture is based on those three layers. The events are sensed and sent to a cloud system server through the gateway (base station). The IoT system consists of a variety of heterogeneous sensors, networks, communication methodologies and processing technologies, but in integrating these different types of technologies, the problem of interoperability arises. To address the problem of interoperability, there must be standardized IoT architecture. (Muhammad et al. 2018).

DOI 10.1201/9781003181545-5

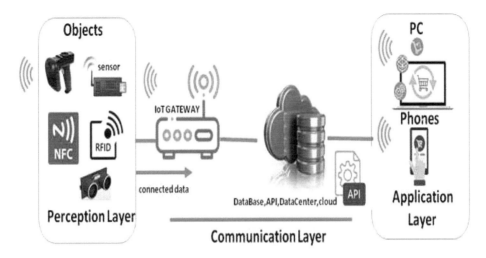

Figure 1. The IoT architecture.

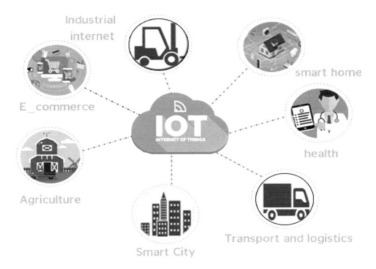

Figure 2. Applications and domains of IoT.

3 IOT APPLICATION DOMAIN

IoT application and domain, has been introduced in many domains such as agriculture, health, transport, e-commerce, smart cities etc. as shown in Figure 2 (Tzafestas 2018).

4 PROPOSED SYSTEM

Internet of Things, or (IoT) is the interconnection between the internet and objects, places and physical environments. The designation designates an increasing number of objects connected to the Internet, thus allowing communication between our so-called physical goods and their digital existence. Over time, the term has evolved and now encompasses the entire ecosystem of connected

Figure 3. The buying process through the IoT and RFID technology.

objects. This ecosystem includes sensor manufacturers, software publishers, incumbent or new operators on the market and integrators. This eclecticism makes it valuable though it's mainly used to compare prices between different websites by recommending the products that are the cheapest and have the best quality for the customer's needs. The customer must choose a suitable product according to his/her purchasing criteria. We will illustrate this part in detail in the next section by proposing a new technique for helping the customer in the selection process. For the time being, we consider organizations an area to implement in the internet of things and suggest a framework for them. Our primary motivation behind integrating the internet of things in this field is to improve the benefit of shopping. The main challenge associated with the internet of things is the organization required to form the network, so we are working on a dynamic processing method that reduces the package size to a large extent which will address the problem of e-commerce. To explain better here is an example: the customer goes to the store and is confused about what he wants to buy, the IoT intervenes here, guiding and facilitating the buying process for the customer through a monitor that shows the description of all products and the best offers available, as shown in the Figure 3.

5 CONCLUSIONS

As a conclusion, this work represents the fundamental aspect to develop an internet of things system (IoT) using RFID technology in e-commerce. Nowadays, internet of things is an innovative technology, which becomes more fundamental to controlling products and services. RFID is not only a technology used for product tagging and tracking, it has great uses like in e-commerce and other applications. However, it is very suitable for companies that need to control products dynamically, to manage data and in surveillance infrastructures. In addition, enterprises who extract useful data from their environment using connected sensors like RFID can develop their business by knowing the needs of their customers without asking them. Therefore, the implantation of IoT system, which integrates RFID sensors, has an important effect in increasing the business of any enterprise. The IoT has many bright points that make the life of the customer easier and help the stores to develop their work. It can provide the knowledge of the geographic position of the nearest stores. Finally as other perspectives on the IOT, we can mention the semantic web and ontology as advantages.

6   ACRONYMS

Table 1.   List of acronyms used in the paper and their full meaning.

Acronyms	Full Meaning
IOT	Internet of things
RFID	Radio Frequency Identification
E-Commerce	electronic commerce
API	Interface Applicative de Programmation
NFC	Near Field Communication

REFERENCES

Hanna Mohammad Said & Abdel-Badeeh M. Salem (2019). Smart E-Business Model based on Block Chain (BC) and Internet of Things (IoT) Technologies, International Journal of Internet of Things and Web Services, Volume 4, ISSN: 2367–9115. https://www.iaras.org/iaras/filedownloads/ijitws/2019/022-0001 (2019).pdf

Hany F. Atlam, & Robert J. Walters. & Gary B. Wills (2018). Internet of Things: State-of-the-art, Challenges, Applications, and Open Issues, International Journal of Intelligent Computing Research (IJICR), Volume 9, Issue 3, 928–938. https://infonomics-society.org/wp-content/uploads/ijicr/published-papers/volume-9-2018/Internet-of-Things-State-of-the-art-Challenges-Applications-and-Open-Issues.pdf

Muhammad B, Rana Asif R, Bilal Khan, Byung-Seo (2018). IoT Elements, Layered Architectures and Security Issues: A Comprehensive Survey Sensors 2018, 18(9), 2796; https://doi.org/10.3390/s18092796

Spyros G. Tzafestas (2018). Ethics and Law in the Internet of Things World, Smart Cities 2018, 1(1), 98–120; https://doi.org/10.3390/smartcities1010006.

Yunusa A & Muhammad Anas G. (2017). IoT on E-commerce, Present and Future: A Review of Alibaba Case Study, Journal of Information Systems Research and Innovation 11(1), 41–46. https://seminar.utmspace.edu.my/jisri/download/Vol11-1/Paper7-Yunusa.pdf

Innovative and Intelligent Technology-Based Services for Smart
Environments – Ben Slama et al (eds)
© 2021 Taylor & Francis Group, London, ISBN 978-1-032-02030-3

Modeling of 3D planar discontinuity in a microstrip patch antenna using the reciprocity theorem combined with MGEC

Raja Mchaalia, Mourad Aidi & Taoufik Aguili
Sys'Com Laboratory ENIT, National Engineering School of Tunis (ENIT),
Tunis-El Manar University, Tunisia

ABSTRACT: Microstrip antennas are widely used in wireless communication, radar, satellite systems, and astronomy applications. Because of their ease of manufacturing, their lightweight, low profile and also their ability to conform to non-planar structures in free space, these structures need tools based on numerical techniques to model and solve some electromagnetic problems like interference and discontinuity. However, the 3D planar discontinuity presents several difficulties related to the volume mesh. Furthermore, the entire volume must be considered, even the smallest details. In this paper, we will develop a formulation based on the hybridization between the reciprocity theorem and the Method of the Generalized Equivalent Circuit (MGEC) to model and analyze a 3D planar discontinuity existing between the excitation source which described by a coaxial cable and a microstrip patch antenna. The values of the current density distribution on either side of the patch antenna and microstrip as well as the relative error value between microstrip patch antenna and excitation source shows the existence of discontinuity.

1 INTRODUCTION

With the rapid development of wireless transmission and mobile networking techniques, various wireless services have emerged and smart devices have become more popular, which has led to an explosive increase in the data traffic of wireless networks (Habaebi et al. 2018, Shandal et al. 2018, Nazeri et al. 2019). Thus, millimeter-wave cellular systems are most likely to be noise limited by interference (Salahat et al. 2017). In the design of millimeter-wave circuits, compensation of microstrip discontinuities is widely used to reduce the effects of discontinuity phenomena (Rappaport et al. 2012). Furthermore, Micro-strip patch receivers (MPAs) have beautiful and boundless features due to their low profile, small size, light weight, low cost as well as the fact they are very simple to construct and are suited to planar and non-planar surfaces. However, their future use in definite systems is finite because of their narrow bandwidth (Ban et al. 2016). Despite their undoubted usefulness, many of these studies have omitted a fundamental yet key feature of the physical signal propagation (Saad & Mohamed 2019). However, signal propagation in wireless channels can be subjected to many types of environmental parameters that degrade its performance. Such factors include noise, discontinuity, interference, large-scale fading, small-scale fading, path loss, delay and other temporal and spectral dynamics of the link that act on the propagated electromagnetic signal in 2-D or 3-D.

Besides, full-wave three-dimensional-discretization numerical methods are considered the most versatile, as they apply to geometrically more complex structures at higher frequencies, just like the high frequencies and short wavelengths of microwave energy make for difficulties in the analysis and design of microwave devices and systems especially in space (3D dimensions). Furthermore, it is necessary to take advantage of the ability to optimize analog and digital techniques simultaneously to reach our goals. There are many algorithms and circuit techniques that have been employed at high frequencies that may bring benefits to the microwave space. For this reason, we resort

DOI 10.1201/9781003181545-6

to electromagnetic methods named "global methods." These methods can be classified into two groups: integral and differential (Mekkioui & Baudrand 2009). Integral methods also adapt well to these kinds of problems. Among the integral methods, we can cite the method of moment (MOM). It gives a general procedure for treating field problems, but the details of the solution vary widely with the particular problem. It is perhaps the most widely used tool for electromagnetic modeling. The advantages of the method of moment are accuracy, versatility, and the ability to compute near as well as as-far zone parameters. The most widely used forms of this method are the thin wire computer programs (Davidson 2010). Despite its advantages, it has been proven in its insufficiency when it comes to 3D problems (Newman & Pozar 1978, Garg 2008). To remedy this drawback, this paper develops an accurate method for analyzing discontinuities in 3D space for microstrip patch antennae. The method presented is based on the integral equation approach. The integral equation is derived through application of the Reciprocity Theorem (RT) (Harrington 1961) and solved by the Method of Moment. To derive a realistic formulation, a coaxial excitation mechanism is used. Overall, to address these inadequacies, we propose a hybridization of methods that combines the one mentioned before described by the reciprocity theorem with the MGEC (Method of Generalized Equivalent Circuits). This hybridization will analyze a discontinuities problem that exists between a microstrip patch antenna in a free space and a coaxial source (Mchaalia et al. 2017, 2019).

2   FORMULATION

In this section, the basic mainstream of the validation process of analytical techniques, which is based on the reciprocity theorem, can be detailed then explored to show how it could be important using such a Reciprocity Theorem (RT) within the electromagnetic field (Mchaalia 2019). After exploration, the formulation of the main problem would be explicitly defined to turn this work into operative balancing behavior of sight satisfaction searches with the required ideas using a digital concept.

2.1   *The general form of the reciprocity theorem*

The reciprocity theorem is among the most useful tools in field and circuit problems. Consider two sets of sources (J^a, M^a) and (J^b, M^b) existing in the same linear medium and at the same frequency. Denote the electromagnetic field as produced by the "a" as being E^a, H^a, and those produced by the "b" sources as E^b, H^b. The Reciprocity theorem formulation in the integral form is given by the below equation which is named the "Lorentz Reciprocity Theorem".

$$\oiint \left(E^a \times H^b - E^b \times H^a \right) dS$$

$$= \iiint \left(E^a \times J^b - H^a \times M^b - E^b \times J^a + H^b \times M^a \right) d\tau \tag{1}$$

2.2   *Reciprocity theorem for microstrip patch antenna*

We consider a structure given by Figure 1. It describes a patch antenna powered by a microstrip and excited by a coaxial cable. This structure is in an open-end homogeneous wave guide with the relative permittivity equal to that of air. The two components of the structure didn't exist in the same plane does it mean that the microstrip exists in the plane orthogonal to the patch antenna plane. In most cases the coaxial feed or "launcher" is designed to allow only transverse electromagnetic (TEM) propagation and the feed's center conductor is small compared to a wavelength. In these cases, the radial electric field will be dominant on the aperture and we can replace the feed by an equivalent magnetic surface current sometimes called a "frill" current and noted M_s. The magnetic current source, M_s, is coupled with the current distributions, J_{s_1}, on the patch antenna and J_{s_2} on the microstrip to produce these couples of electric and magnetic fields $\{E^t, H^t\}$.

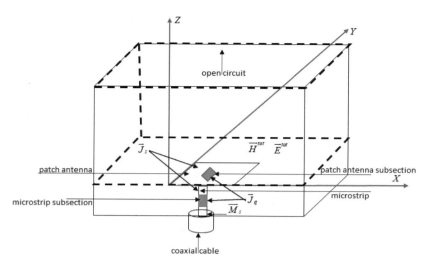

Figure 1. Structure design: a = 47.55mm, b = 22.15, c = 15.5mm, L = 24.5mm, W = 2.8mm, l = 1.5mm, h = 1.5mm, r_in = 0.37mm, r_out = 0.8mm, $\varepsilon_r = 1$.

We applied the reciprocity theorem to the structure of Figure 1, we obtained the following equation:

$$\iiint_V \left(E^t, J_q\right) \cdot dV = \iiint_V \left(E_q \cdot J_s - H_q \cdot M_s\right) \cdot dV \tag{2}$$

Where V represents the waveguide interior volume, E_q and H_q represent the electric and magnetic field respectively associated with the test current, and J_q is a weighting function in the Method of Moment (MOM) of the patch antenna and the microstrip.

$E^t \cdot J_q$ is equal to zero because J_q exists in a subsection of a planar structure and E^t occupied the hole volume of waveguide. The left-hand side of the equation (2) vanishes. In fact, we can reduce the remaining integral in (2) to the surface integrals which results in:

$$\iint_{S_1} \left(\left(\vec{E}_{q_1} + \vec{E}_{q_2}\right)\Big|_{z=0} \cdot \vec{J}_{s_1} \right) \cdot dS_1 + \iint_{S_2} \left(\left(\vec{E}_{q_1} + \vec{E}_{q_2}\right)\Big|_{y=0} \cdot \vec{J}_{s_2} \right) \cdot dS_2$$

$$= \iint_{S_{feed}} \left(\vec{H}_{q_1} \cdot \vec{M}_s\right) \cdot dS_{feed} + \iint_{S_{feed}} \left(\vec{H}_{q_2} \cdot \vec{M}_s\right) \cdot dS_{feed} \tag{3}$$

where $dS_1 = dxdy$ (antenna plane), $dS_2 = dxdz$ (microstrip plane), and the current density distribution of the patch antenna and the microstrip are both given by the following equation:

$$\begin{cases} \vec{J}_{s_1}(x,y) = x_{p_1} g_{p_1}(x,y)\, \vec{y} \\ \vec{J}_{s_2}(x,z) = x_{p_2} g_{p_2}(x,z)\, \vec{z} \end{cases} \tag{4}$$

Equation 4 has the following matrix form which is written as:

$$I_1 M_{11} X_1 + I_2 M_{21} X_1 + I_1 M_{12} X_2 + I_2 M_{22} X_2 = I_1 A_1 + I_2 A_2 \tag{5}$$

The matrix resolution, which given by equation 5, will give the value of the current density distribution in the patch antenna and is expressed by:

$$J_y(x,y) = \left(\begin{array}{c} \sum\limits_{p,q} \left(\sum\limits_{m,n} I_p \langle g_p(x,y), f_{mn} \rangle f_{mn} \right)^{TE} \\ + \sum \left(\sum\limits_{m,n} I_p \langle g_p(x,y), f_{mn} \rangle f_{mn} \right)^{TM} \end{array} \right) \cdot \vec{x} \tag{6}$$

The same current expression is used in the microstrip. All the obtained results will be discussed in the next section (Figure 2).

3 RESULTS AND DISCUSSION

The theoretical method developed above has been implemented on MATLAB for calculating the current density distribution of the microstrip patch antenna excited by a coaxial cable. The dimensions of antenna are chosen to ameliorate its characteristic.

The current behavior in the patch antenna (Figure 4) verifies the boundary conditions of the chosen test functions at a frequency equal to 28 GHz. In fact, the current density distribution values for both the microstrip and patch antenna are taken for a test functions number of P=Q=5 and for that mode, a functions number equal to M=N=350.

We noticed in the patch antenna current figure $x = L/2$ (L represents microstrip patch antenna length), the area where the patch antenna is connected to the microstrip, that the current is not uniform and its value is very small. This is explained by the existing of disturbances called discontinuities coming from the microstrip via the excitation source. We noticed also that the current behavior presents two lobes which are symmetrical on both sides of a source position (Figure 4). Figure 3 shows the current density distribution of the microstrip component which is the maximum for $z = l/3$ (l is the microstrip length) where it is connected to the patch antenna. Hence, the relative error between two components regarding the microstrip and patch antenna is equal to 0.4%. Overall, the 3D planar discontinuity behavior that mentioned in the patch antenna graph described the specific dimensions of the microstrip that taken in the above structure. But, if we increase or

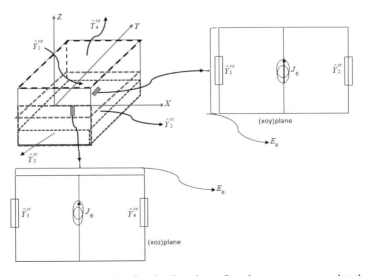

Figure 2.   Generalized equivalent circuit of each subsections of patch antenna connected to the strip.

37

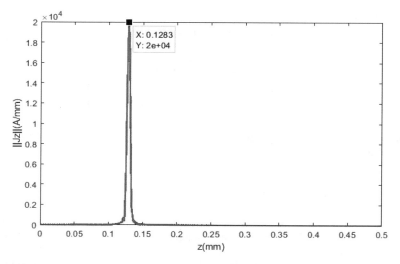

Figure 3. Value of current density distribution in the microstrip.

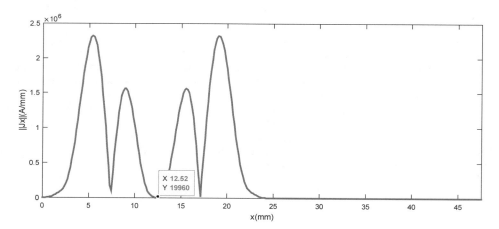

Figure 4. Value of current density distribution of the patch antenna.

decrease the microstrip length, the antenna current density value changes. As a result, the patch antenna discontinuity affects with microstrip dimensions.

4   CONCLUSION

In this work, we modeled a microstrip antenna in 3D space excited by coaxial cable via a microstrip line. The hybridization principle that combined the reciprocity theorem with MGEC to develop the current density expression was used. The obtained theoretical results of the current density distribution show the existence of discontinuity between the patch antenna element and the microstrip via a source. According to the relative error value of both the patch antenna and microstrip proves that the discontinuity between the excitation source and radiation elements in free space exist. However, in future works, we will implement the discontinuity circuit between the mentioned two elements for 5G antenna networks, calculating the scattering parameters and radiation pattern of the microstrip patch antenna.

REFERENCES

Ban, Y.L., Li, C., Wu, G. and Wong, K.L., 2016. 4G/5G multiple antennas for future multi-mode smartphone applications. *IEEE access*, *4*, pp. 2981–2988.

Davidson, D.B., 2010. *Computational electromagnetics for RF and microwave engineering*. Cambridge University Press.

Garg, R., 2008. *Analytical and computational methods in electromagnetics*. Artech house.

Habaebi, M.H., Janat, M. and Islam, M.R., 2018. Beam steering antenna array for 5G telecommunication systems applications. *Progress in Electromagnetics Research*, *67*, pp. 197–207.

Harrington, RF. Time-harmonic electromagnetic fields. 1961, *McGraw-Hill*.

Mchaalia, R., 2019. *Planar Discontinuity Modeling in 3D Space by Application of Reciprocity Theorem Combined with MGEC* (Doctoral dissertation).

Mchaalia, R., Aidi, M. and Aguili, T., 2017, August. A new 3D MOM-GEC formulation based on reciprocity theorem: Analysis of the dipole antenna. In *2017 International Applied Computational Electromagnetics Society Symposium (ACES)* (pp. 1–2). IEEE.

Mchaalia, R., Aidi, M. and Aguili, T., 2019. Study of planar structure applying reciprocity technique combined with MGEC and analysis of discontinuity. *International Journal of RF and Microwave Computer-Aided Engineering*, *29*(9), p.e21856.

Mekkioui, Z. and Baudrand, H., 2009. Bi-dimensional bi-periodic centred-fed microstrip leaky-wave antenna analysis by a source modal decomposition in spectral domain. *IET microwaves, antennas & propagation*, *3*(7), pp. 1141–1149.

Nazeri, A., Abdolali, A. and Mehdi, M., 2019. An Extremely Safe Low-SAR Antenna with Study of Its Electromagnetic Biological Effects on Human Head. *Wireless Personal Communications*, *109*(2), pp. 1449–1462.

Newman, E. and Pozar, D., 1978. Electromagnetic modeling of composite wire and surface geometries. *IEEE Transactions on Antennas and Propagation*, *26*(6), pp. 784–789.

Rappaport, T.S., Gutierrez, F., Ben-Dor, E., Murdock, J.N., Qiao, Y. and Tamir, J.I., 2012. Broadband millimeter-wave propagation measurements and models using adaptive-beam antennas for outdoor urban cellular communications. *IEEE transactions on antennas and propagation*, *61*(4), pp. 1850–1859.

Saad, A.A.R. and Mohamed, H.A., 2019. Printed millimeter-wave MIMO-based slot antenna arrays for 5G networks. *AEU-International Journal of Electronics and Communications*, *99*, pp. 59–69.

Salahat, E., Kulaib, A., Ali, N. and Shubair, R., 2017, June. Exploring symmetry in wireless propagation channels. In *2017 European Conference on Networks and Communications (EuCNC)* (pp. 1–6). IEEE.

Shandal, S.A., Mezaal, Y.S., Mosleh, M.F. and Kadim, M.A., 2018. Miniaturized Wideband Microstrip Antenna for Recent Wireless Applications. *Advanced Electromagnetics*, *7*(5), pp. 7–13.

Innovative and Intelligent Technology-Based Services for Smart Environments – Ben Slama et al (eds)
© 2021 Taylor & Francis Group, London, ISBN 978-1-032-02030-3

Convolutional Neural Networks based decoding in Sparse Code Multiple Access for different channel models

M. Hizem, I. Abidi & R. Bouallegue
Tunisia-Innov'COM, SUP'COM, University of Carthage, Tunisia

ABSTRACT: In this paper, the performance of Convolutional Neural Networks (CNN) based decoding in Sparse Code Multiple Access (SCMA) for different channel models is investigated. SCMA gets more attention in the research field to meet the requirements of the next generation of wireless communication networks. It is certainly considered as a possible candidate for next-generation systems that can achieve better performance in terms of spectral efficiency and latency and solve the problem of massive connectivity. SCMA encoding is studied in order to understand the architecture of our system. A blind decoding based on CNN is described and used in order to minimize the very high detection complexity, which is the biggest challenge of SCMA. Then, another channel model is employed in addition to AWGN. Through simulations, we showed that the use of the Rayleigh channel affects the BER performance compared to the Gaussian one.

1 INTRODUCTION

There are several requirements for the next generation of wireless communication networks. Massive connectivity, higher spectral efficiency, better quality of service and lower latency are some of the exigencies that should be satisfied with new waveform and access designs (Andrews et al. 2014, Li et al. 2014). Next, wireless networks should perform well to support much larger capacity and more options for more connected mobile devices to meet with emergent applications such as the Internet of Things (IoT). The future generation is essential to deliver maximum data bit rates and high spectral efficiency in an extended coverage area (Ding et al. 2017). Non-Orthogonal Multiple Access (NOMA) schemes have been proposed to enhance the multiple access gain and increase the number of users (Saito et al. 2013, Wang et al. 2015). There are two types of NOMA systems: code-domain multiplexing and power-domain multiplexing. SCMA is a scheme of the first type and a potential candidate of NOMA for future generations of wireless communication networks (Nikopour & Baligh 2013). SCMA is a waveform in which input bits are directly mapped to multidimensional complex codewords selected from a predefined codebook set (Taherzadeh et al. 2014). On the other hand, in SCMA, the spreading matrix is required to be sparse such that only a small number of users overlap in each shared resource in order to minimize the multiuser interference (Vameghestahbanati et al. 2017).

The performance of SCMA systems can be improved by reducing the very high detection complexity. For that, some methods have been developed in recent research, for example the low-density parity check (LDPC) message passing algorithm detector (LDPC-MPA) studied in Xiao et al. 2015, the SCMA decoder exploiting a low complexity maximum-likelihood approximation proposed in Alizadeh & Savaria 2016, and the Max-log Message Passing Algorithm (MPA) detector presented in Liu et al. 2016. Despite the fact that these methods have high computational overheads that limit the real-time operation of SCMA, they have even managed to provide a good compromise between Bit Error Rate (BER) performance and computational complexity (Alam & Zhang 2017).

Recently, different neural network technologies have been used to successfully resolve the problems of classification and image recognition. As in Kim et al. 2018, the application of neural

DOI 10.1201/9781003181545-7

networks with SCMA detection allows us to obtain a low complexity SCMA blind detector with good performance. In (Abidi et al. 2019), we proposed a new convolutional neural network based decoding system for SCMA. Thanks to its blind decoding strategy, our decoder outperforms the limitations of traditional deep neural networks. The simulation results show the performance gain in terms of both BER and computational complexity in comparison with the decoding strategy based on the iterative approach MPA. In this paper, we extend the work in Abidi et al. 2019 developed in the context of a Gaussian channel to other more realistic channel models in order to evaluate their effect on the performance of our system. Simulations have shown that the use of the Rayleigh channel affects the BER performance compared to the Gaussian one.

The rest of this paper is organized as follows: in section 2, we present the system model of SCMA and our proposed convolutional neural network based decoder, the simulation results are given in section 3, and we conclude and suggest perspectives in section 4.

2 SYSTEM MODEL FOR SCMA

The system model of SCMA is shown in Figure 1. We consider a classic system model with J-users spreading over K-resources through different model channels (AWGN and Rayleigh). We assume that K is lower than J and that the ration J/K is noted as the overloading factor. In SCMA, incoming bit streams are directly mapped to multidimensional complex codewords chosen from a predefined codebook, where each codeword represents a spread transmission layer. In the following passage, the encoding and detection procedures are explained in more detail (Abidi et al. 2019).

2.1 *SCMA encoder*

An SCMA encoder is a mapping from $\log_2(M)$ bits to a K-dimensional codebook of size M (Kschischang et al. 2001). The K-dimensional complex codewords of the codebook are sparse vectors with J non-zero entries. SCMA codebooks are elaborated to decrease the detection complexity. The codewords are constituted of complex multidimensional symbols, and have the same sparse pattern if they are in the same codebook.

The SCMA transceiver can be defined as a mapping of coded bits b_j to a multidimensional codeword x_j. Thereafter, the data is modulated to the subcarriers. The received signal can be written as follows:

$$y = \sum_{j=1}^{J} diag\left(h_j\right) x_j + w \tag{1}$$

Where $x_j = (x_{1j}, \ldots, x_{Kj})^T$ is the SCMA codeword of layer j, $h_j = (h_{1j}, \ldots, h_{Kj})$ is the channel vector of layer j, w is Gaussian with zero mean and variance $N_0 I$, is the ambient noise.

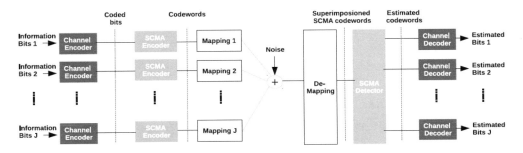

Figure 1. Block diagram of SCMA system model.

2.2 *Proposed convolutional neural networks based decoder*

This section proposes a blind decoder for SCMA based on CNNs (Krizhevsky et al. 2012). This detector uses a variation of multilayer perceptions concepted to need minimal preprocessing. These deep-learning systems realize an automatic feature extraction without any human contribution, contrary to most traditional machine-learning networks. The deep learning's aptitude to treat and learn from enormous quantities of unlabeled data, allow it a have particular advantage over previous algorithms (Lawrence et al. 1997). In these networks, each layer trains on a separate set of characteristics founded on the previous layer's output (Kim et al. 2018). Enjoying the advantages of its structure, the CNNs may perfectly exceed the constraints of traditional neural networks. In fact, a CNN decomposes into a number of layers such as non-linearity, convolutional and fully connected layers.

In adopting CNNs with SCMA, our purpose is to build a blind decoding approach with high performance and low complexity. This latter objective can be evaluated by determining the values of mean square error (MSE) and weight updates, which has been done in Abidi et al. 2019.

In order to elaborate our model, it is necessary to specify the number of layers, the learning weights, the number and size of filters, and other adjusted parameters. These parameters, along with the network architecture, are initialized at the start and do not change during the training process, except the values of the filter matrix and connection weights which will be updated.

For the implementation, we structured first the data for training and testing. Thereafter, we constructed the layers and the filters to produce a suitable model, which will be utilized to detect blindly the codewords forwarded by each user, without preliminary information of signal and channel characteristics.

3 SIMULATION RESULTS

Simulation results show that the CNNs for blind decoding achieve better performance than MPA in terms of both BER and computational complexity. This is explained by the fact that all the stages of optimizing the weights and adjusting the system parameters are achieved during the training process. Furthermore, Figure 2a demonstrates that more iterations are required to reach a better performance by using MPA, which generates an increase in computation time. We noticed that the computational complexity of our proposed method is much lower than the MPA based decoder scheme. This is shown in Figure 2b, from which we can deduce that the computation time of our proposed method is lower than the MPA one even with one iteration.

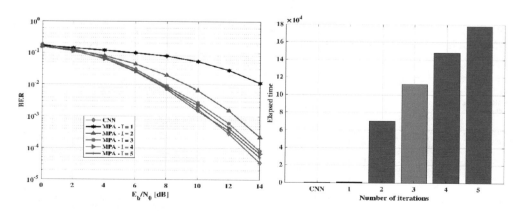

Figure 2.   BER performance of the MPA based decoder vs CNNs based decoder for SCMA (Abidi et al. 2019).

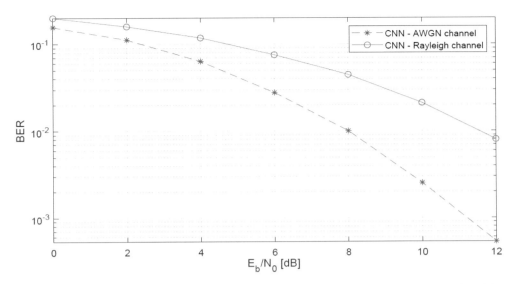

Figure 3. BER performance of CNNs based decoding in SCMA for AWGN and Rayleigh channel models.

In Figure 2, simulations were done in the context of an Additive White Gaussian Noise (AWGN) channel. In order to evaluate the effect of more realistic channel models on the performance of our system, we extended our work to the case of a Rayleigh channel, as shown in Figure 3. We can see that this extension affected the BER performance of our network compared to AWGN, which was predictable due to the multi-path effect added by Rayleigh channel.

4 CONCLUSION

In this paper, we have developed a new approach of decoding for SCMA based on CNNs. The simulation results demonstrated that our proposed method outperforms the MPA scheme in terms of both BER and computational complexity. Moreover, we showed that the performance is affected using the Rayleigh channel. It is for this reason that future research will focus on how to improve the performance of our system in this case. A channel estimation or further improvements on both transceivers and receivers could be explored.

REFERENCES

Abidi I., Hizem M., Ahrizz I., Cherif M. & Bouallegue R. 2019. Convolutional neural networks for blind decoding in sparse code multiple access, Accepted in 15th International Wireless Communications & Mobile Computing Conference (IWCMC).

Alam M. & Zhang Q. 2017. Performance study of scma codebook design, Proc. IEEE Wireless Communications and Networking Conference (WCNC), pp. 1–5.

Alizadeh R. & Savaria Y. 2016. Performance analysis of a reduced complexity scma decoder exploiting a low-complexity maximum-likelihood approximation, Proc. IEEE International Conference on Electronics, Circuits and Systems (ICECS), pp. 253–256.

Andrews J., Buzzi S., Choi W., Hanly S., Lozano A., Soong A. & J. Zhang 2014. What will 5G be?, IEEE J. Sel. Areas Commun, vol. 32, no. 6, pp. 1065–1082.

Ding Z., Liu Y., Choi J., Sun Q., Elkashlan M., Chih-Lin I. & Vincent Poor H. 2017. Application of non-orthogonal multiple access in lte and 5g networks, IEEE Communications Magazine, vol. 55, no. 2, pp. 185–191.

Kim M., Kim N., Lee W. & Cho D. 2018. Deep learning-aided scma, IEEE Communications Letters, vol. 22, no. 4, pp. 720–723.

Krizhevsky A., Sutskever I. & Hinton G. E. 2012. ImageNet classification with deep convolutional neural networks, In Advances in Neural Information Processing Systems, pp. 1097–1105.

Lawrence S., Giles C. L., Tsoi A. C. & Back A. D. 1997. Face recognition: A convolutional neural-network approach, IEEE Transactions on neural networks, vol. 8, no. 1, pp. 98–113.

Kschischang F. R., Frey B. J. & Loeliger H-A 2001. Factor graphs and the sum-product algorithm, IEEE Transactions on information theory, vol. 47, no. 12, pp. 498–519.

Li Q., Niu H., Papathanassiou A. & Wu G. 2014. 5G network capacity: key elements and technologies, IEEE Vehicular Tech. Mag, vol. 9, no. 1, pp. 71–78.

Liu J., Wu G., Li S. & Tirkkonen O. 2016. On fixed-point implementation of log-mpa for scma signals, IEEE Wireless Communications Letters, vol. 5, no. 3, pp. 324–327.

Nikopour H. & Baligh H. 2013. Sparse Code Multiple Access, Proc. IEEE International Conference on Personal Indoor and Mobile Radio Communications (PIMRC), pp. 332–336.

Saito Y., Kishiyama Y., Benjebbour A., Nakamura T., Li A. & Higuchi K. 2013. Non-orthogonal multiple access (noma) for cellular future radio access, Proc. IEEE 77th Vehicular Technology Conference (VTC Spring), pp. 1–5.

Taherzadeh M., Nikopour H., Bayesteh A. & Baligh H. 2014. SCMA Codebook Design, Proc. 2014 IEEE 80th Vehicular Technology Conference (VTC2014-Fall).

Vameghestahbanati M., Marsland I., Gohary R. H. & Yanikomeroglu H. 2017. Polar Codes for SCMA Systems, Proc. IEEE 86th Vehicular Technology Conference (VTC-Fall).

Wang B., Wang K., Lu Z., Xie T. & Quan J. 2015. Comparison study of non-orthogonal multiple access schemes for 5g, Proc. IEEE International Symposium on Broadband Multimedia Systems and Broadcasting (BMSB), pp. 1–5.

Xiao B., Xiao K., Zhang S., Chen Z., Xia B. & Liu H. 2015. Iterative detection and decoding for SCMA systems with LDPC codes, Proc. International Conference on Wireless Communications and Signal Processing (WCSP), pp. 1–5.

Innovative and Intelligent Technology-Based Services for Smart Environments – Ben Slama et al (eds)
© 2021 Taylor & Francis Group, London, ISBN 978-1-032-02030-3

Electromagnetic near-field study of electric probes for EMC applications

Intissar Krimi
Higher Institute of Computer Science and Multimedia, University of Gabes, Tunisia

Sofiane Ben Mbarek
Physical Science and Engineering Division, King Abdullah University of Science and Technology, Thuwal, Saudi Arabia

Hechmi Hattab
Higher Institute of Computer Science and Multimedia, University of Gabes, Tunisia

Fethi Choubani
Innov'Com Laboratory, SUPCOM, University of Carthage, Tunisia

ABSTRACT: In this work, we present a new modeling approach which can be a useful tool to design electric probes for Electromagnetic Compatibility applications. Our approach is based on the Finite Difference Time Domain method and spline interpolation. The analysis of the results of the near-electric field maps from the FDTD and spline method, shows that our method could be used to enhance the spatial resolution accurately.

Keywords: CPW probe, Near Field, FDTD, Spline interpolation.

1   INTRODUCTION

Nowadays, electronic systems are integrating smaller and more sophisticated features in a confined space. Quite often, some circuits generate Electromagnetic Interference (EMI) problems. It is, therefore, essential to understand the Electromagnetic (EM) environment (Liu et al. 2008). The characterization of EMI problems using Near-Field (NF) measurements is necessary. The use of NF techniques in Electromagnetic Compatibility (EMC) applications increases rapidly in power electronics (Ayestaràn & Las-Heras 2006). These techniques constitute an efficient approach to characterizing complex radiating systems (Ben Mbarek et al. 2017), (Baudry et al. 2006). Ideally, the NF measurements should be accurate, however that is not possible due to the dimensions of the probe (Ben Mbarek et al. 2018). For the conception of those probes, it's important to have more information about their spatial resolution and sensitivity with high accuracy.

In this paper, we present a new modeling approach which could be a useful tool for the design of electric probes for EMC applications. Our approach is based on the Finite Difference Time Domain (FDTD) method to which we add quadratic and cubic splines interpolation. This will conciliate the accuracy of a small probe and the sensitivity of a large one. In the first part, we present the mathematical equations and boundary conditions used in the development of our FDTD model with splines interpolation. In the second part, the results and the error analysis are presented to validate the accuracy of our proposed method.

2   MATHEMATICAL BACKGROUND

Before the examination of the interpolation method, we shall start with an FDTD formulation for a two-dimensional Coplanar Waveguide (CPW) probe (Ben Hassen et al. 2016; Ben Mbarek et al.

DOI 10.1201/9781003181545-8

2017). FDTD is one of the most common computational EM methods. Space and time are divided into Yee Cells (Yee 1966). Maxwell's equations in time domain and in general form are converted to a FDTD equivalent problem with the Yee algorithm (Yee 1966). According to the boundary conditions in the CPW waveguide, the discrete Maxwell's equations are in the following form:

$$\frac{\partial D_z}{\partial t} = \frac{1}{\sqrt{\mu_0 \varepsilon_0}} \left(\frac{\partial H_y}{\partial x} - \frac{\partial H_x}{\partial y} \right) \tag{1}$$

$$\frac{\partial H_y}{\partial t} = \frac{1}{\sqrt{\mu_0 \varepsilon_0}} \left(\frac{\partial E_z}{\partial x} \right) \tag{2}$$

$$\frac{\partial H_x}{\partial t} = -\frac{1}{\sqrt{\mu_0 \varepsilon_0}} \left(\frac{\partial E_z}{\partial y} \right) \tag{3}$$

Where μ, ε and D are magnetic permeability, electric permittivity and the electric displacement field respectively. These equations are expressed by means of central finite differencing. A sinusoidal pulse acted as an excitation wave with a frequency f of 3 GHz.

$$E_z = 1 * (\sin (Tw\Delta t)) \tag{4}$$

Where w is the angular frequency, T is number of time steps and $\Delta t = \Delta x = 2c$. The FDTD domain environment was surrounded by the Perfectly Matched Layer (PML) with a thickness of 10 mm to prevent the reaction at the boundary of the environment. In this work, we will approximate the smooth curves through FDTD points by piecewise polynomial interpolation using the spline function (Kharab & Guenther 2018; Warbhe & Gomes 2016; Usman & Ramdhani 2019).

Quadratic interpolation gives a better fit to a continuous function than a linear spline interpolation does. Let f be a function defined on the interval [a, b] and $a = x_0 < x_1 < \ldots < x_k = b$ are $k + 1$ distinct points (nodes) in [a,b]. We consider the piecewise (spline) quadratic and cubic interpolation of f at those nodes. Whereas the quadratic spline interpolation requires the continuity of first derivatives, at the interpolating nodes and within the intervals, the cubic one requires more over second derivatives. This gives a smoother function in the cubic interpolation than the quadratic case. The continuity of the 1st derivative means that the function $S(x) = f$ will not have sharp corners, and the continuity of the 2nd derivative means that the radius, of curvature, is defined at each point. The equation of the cubic spline in the sub-interval is $[x_{i-1}; x_i]$, such that:

$$s_i (x) = a_i x^3 + b_i x^2 + c_i x + d_i; \qquad for\ i = 1, 2, \ldots, k. \tag{5}$$

where a_i, b_i, c_i and d_i are the $4k$ coefficients. Whereas for the quadratic case we have:

$$s_i (x) = a_i x^2 + b_i x + c_i; \qquad for\ i = 1, 2, \ldots, k. \tag{6}$$

where a_i, b_i, and c_i are the $3k$ coefficients.

In the next part, we will present the results of the FDTD modeling using quadratic and cubic splines interpolation.

3 RESULTS AND DISCUSSION

Figure 1 presents the distribution of the E_z component in the xy plane of our electrical CPW probe and the incident and transmitted E_z along the y axis. The maximum of E_z is concentrated at the tip. The spatial resolution of the probe will be related to the tip size and the sensitivity of the CPW probe and will depend on electric intensity at the probe's tip. The Figure 1 corresponds to the 2D FDTD result of the CPW probe. The comparison between the spatial distribution of E_z obtained by quadratic and cubic splines interpolation is shown in Figure 2 and Figure 3. In Figure 2-a and

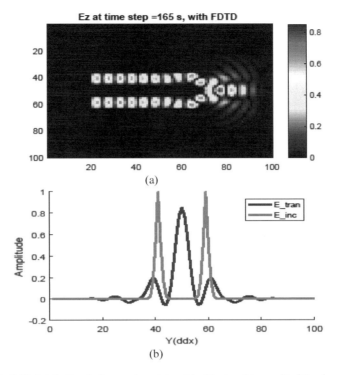

Figure 1.   The E_z field distribution in the xy plane (a) and incident and transmitted E_z along y axes (b).

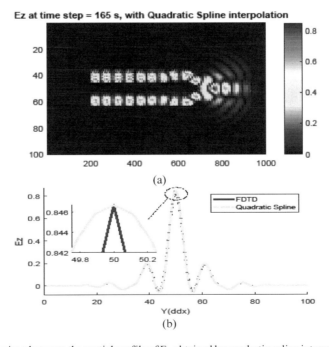

Figure 2.   Comparison between the spatial profile of E_z obtained by quadratic spline interpolation and FDTD: (a) the E-field distribution in the xy plane (b) E_z at the tip of the CPW probe.

Figure 3-a, the results of the distribution of E_z in the xy using FDTD with quadratic and cubic splines interpolation respectively are shown. In Figure 2-b and Figure 3-b, the electric field at the tip of CPW probe using FDTD with quadratic and cubic splines interpolation respectively is shown. We notice, from the spatial distribution of the E-field, that the cubic and quadratic splines function gives a better resolution than FDTD one. The influence of polynomial degree and the smoothing parameter is depicted in Figure 4. We can see the spatial profile obtained by quadratic spline interpolation gives a wider lobe than the profile obtained by cubic spline interpolation. This is due to the number of unknowns in the polynomial. In other words, cubic spline interpolation offers a more exact approximation than quadratic spline interpolation. Indeed, more of the degree of the spline is higher and more the approximation is closer to FDTD data. Thus, to improve the approximation we increase the spline numbers and not the degree of the polynomial.

In order to analysis the proposed method, the E-field E_z according to the y axis obtained by FDTD (Table 1) was adjusted using the splines. Here we apply quadratic and cubic splines interpolation on the intervals [48, 49], [49, 50], and [50,51]. These splines are given, respectively, as follows:

$$Q_i(y) = a_i(y - y_i)^2 + b_i(y - y_i) + c_i \quad i = 1, \ldots, k. \tag{7}$$

$$C_i(y) = a_i(y - y_i)^3 + b_i(y - y_i)^2 + c_i(y - y_i) + d_i \quad i = 1, \ldots, k. \tag{8}$$

To find a smoother interpolating function to represent a better spatial resolution, we use parametric smoothing $= 10$. This parametric can give us 10 data points between one sub interval of electric, while FDTD uses only 2 data points with straight lines. A summary of the calculation

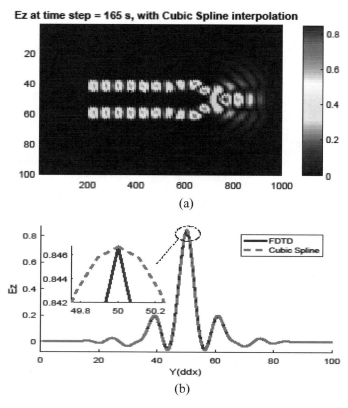

(a)

(b)

Figure 3. Comparison between the spatial profile of E_z obtained by cubic spline interpolation and FDTD: (a) the E-field distribution in the xy plane (b) E_z at the tip of the CPW probe.

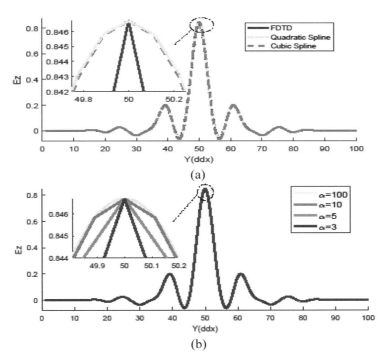

Figure 4. E_z obtained by cubic and quadratic splines interpolation (a) Influence of the smoothing parameter (b).

Table 1. Maximum of E_z in sub-interval [48,51] obtained by FDTD.

y_i	48	49	50	51
E_z	0.5918	0.7758	0.8464	0.7758

Table 2. Electric intensity E_z obtained by FDTD and interpolation.

y	Q_y	C_y	E_z
48.4444	0.6840	0.6836	0.6736
48.6667	0.7238	0.7241	0.7145
49.5556	0.8327	0.8321	0.8152
50.5556	0.8248	0.8241	0.8073

results is given in the (Table 2). Where Q(y) is the piecewise quadratic, C(y) is the picewise cubic and E_z is the computed field from FDTD.

In order to analyze the efficiency of the proposed method, the error analysis for the electric field E_z was calculated as follows:

$$RelativeError = \frac{|E_{zi}(y) - E_z(y)|}{E_z(y)} \qquad (9)$$

Table 3. Error analysis for the electric field E_z.

	FDTD with quadratic spline	FDTD with cubic spline
Maximum amplitude peak of E_z V/m)	0.847	0.847
Relative error (%)	1.09	1.06
MSE	8.7699×10^{-6}	8.2242×10^{-6}

$$MSE = \frac{1}{n}\sum_{i+1}^{n}|E_{zi}(y) - E_z(y)|^2 \qquad (10)$$

Where $E_z(y)$ is the electric field in FDTD, $E_{zi}(y)$ is the electric field interpolated with cubic or quadratic spline, and n is the total number of time steps. Table 3 presents the error analysis for the electric field E_z at the tip of the CPW probe. We can notice that the cubic spline shows lower relative error and lower MSE than the quadratic spline interpolation.

4 CONCLUSION

The FDTD method with splines interpolation was applied to analyzing the Near-Field of Electric Probes for Electromagnetic Compatibility (EMC) applications. The Electric Field analysis proved that the proposed method was able to obtain a good spatial resolution. The error analysis was investigated. The results show that our approach can be a promising method for the conception of electric probes for electromagnetic near-field applications. For future works, more simulations will be done to improve the sensitivity of the CPW probes.

REFERENCES

Ayestaràn, R. G., & Las-Heras, F. (2006). Near field to far field transformation using neural networks and source reconstruction. *Journal of Electromagnetic Waves and Applications, 20(15),* 2201–2213.

Baudry, D., Louis, A., & Mazari, B. (2006). Characterization of the open-ended coaxial probe used for near-field measurements in EMC applications. *Progress In Electromagnetics Research, 60,* 311–333.

Ben Hassen, M., Mbarek, S. B., & Choubani, F. (2016, December). 2-D FDTD analysis of CPW antenna for electromagnetic near-field applications. *In 2016 7th International Conference on Sciences of Electronics, Technologies of Information and Telecommunications (SETIT) (pp. 62–65). IEEE.*

Ben Mbarek, S., Choubani, F., & Cretin, B. (2017, March). Near-field microwave CPW antenna for scanning microscopy. *In 2017 11th European Conference on Antennas and Propagation (EUCAP) (pp. 2812–2815). IEEE.*

Ben Mbarek, S., Choubani, F., & Cretin, B. (2018). Investigation of new micromachined coplanar probe for near-field microwave microscopy. *Microsystem Technologies, 24(7),* 2887–2893.

Kharab, A., & Guenther, R. (2018). An introduction to numerical methods: a MATLAB® approach. *CRC press.*

Liu, B., Beghou, L., Pichon, L., & Costa, F. (2008). Adaptive genetic algorithm based source identification with near-field scanning method. *Progress In Electromagnetics Research, 9,* 215–230.

Usman, K., & Ramdhani, M. (2019, July). Comparison of Classical Interpolation Methods and Compres- sive Sensing for Missing Data Reconstruction. *In 2019 IEEE International Conference on Signals and Systems (ICSigSys) (pp. 29–33). IEEE.*

Warbhe, S., & Gomes, J. (2016). Interpolation technique using non-linear partial differential equation with edge directed bi-cubic. *International Journal of Image Processing (IJIP), 10(4),* 205.

Yee, K. (1966). Numerical solution of initial boundary value problems involving Maxwell's equations in isotropic media. *IEEE Transactions on antennas and propagation, 14(3),* 302–307.

Innovative and Intelligent Technology-Based Services for Smart Environments – Ben Slama et al (eds)
© 2021 Taylor & Francis Group, London, ISBN 978-1-032-02030-3

Towards a new model of human tissues for 5G and beyond

A. Ben Saada
Innov'COM, SUP'COM, University of Carthage, Tunis, Tunisia

S. Ben Mbarek
Physical Science and Engineering Division, King Abdullah University of Science and Technology, Thuwal, Saudi Arabia

F. Choubani
Innov'COM, SUP'COM, University of Carthage, Tunis, Tunisia

ABSTRACT: A planar multilayered model is investigated in this paper using a cylinder insertion simulating a hair shaft. This model has been proved sensible to the mean hair density and length of the human hair.

1   INTRODUCTION

The unstoppable increase of high data-rate applications in telecommunication has led to a use of wider frequency bands. This expansion in the spectrum has logically found its natural place in the extreme right side of the spectrum, because of the reduced number of allocated applications operating in these frequencies. This licensed frequency jump must prove it is not hazardous to biological tissue. The Active Specific Absorption Rate (SAR) standard using the Finite-Difference-Time-Domaine (FDTD) method (IEC 2017) or Finite-Element-Method (FEM) sets a maximum frequency of 6 GHz, which is far away from the new frequencies allocated for 5G in the K-band and V-band. In this case, analytic estimation of average SAR values may become a good estimation tool. In this paper, we will consider additional sub-components of the human tissue and varying their parameters to see how it can affect the E-field transmitted in the inner layers of the proposed model. In the first section, we will introduce the changes made on the planar multilayered Model and the method used, and then we will introduce the E-field curves in the results section in order to give a in-depth explanation of the plots in the Discussion.

2   MODEL AND METHODS

The model proposed in this paper inherits from the human body its anatomic hierarchy, and the electrical properties of the major components of the human tissues. The model presented in Figure 1 is seen as an enhanced planar and multilayered abstraction of human tissue. Multilayered models have been widely used since the early works on Dosimetry and SAR (Ho et al. 1971; Massoudi et al. 1979; Joines et al. 1974), and in more recent works (Saada et al. 2019; Messaoudi et al. 2018). In this paper, we will show how the planar model is in the edge of validity when it comes to SAR estimation for 5G applications, and especially for enhanced Mobile Broadband (eMBB) usage.

The planar multilayered used in a previous work (Saada et al. 2019) is a simple abstraction of the dominant components of the human body (skin, fat, muscle, bone, etc.) with a thickness depending on the region of body. The new model of Figure 1 is a unit cell including one dielectric cylinder simulating hair shaft, and three dielectric slabs constituted consecutively by skin, fat and muscle. The dielectric cylinder starting from the bottom of the skin and ending in the air simulates one

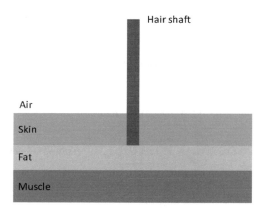

Figure 1. Planar, multilayered model with a finite cylindrical inclusion.

hair follicle without a sebaceous gland. The hair follicle is reduced in a hair shaft; all surrounding elements are neglected in this paper (Blume-Peytavi et al. 2008). Despite the tapered shape of the reel hair shaft, we will only consider a constant circular cross-section.

The electric properties of the used tissues are available on different Electronic Design-Automation (EDA) tools and internet databases (Andreuccetti et al. 2007; Sasaki et al. 2014). However, the electric properties of the hair shaft are unavailable in a wide range of the spectrum. Only some few experimental values of the permittivity are obtained for frequencies below 5 GHz (You et al. 2015).To overcome this limit, we can rely on the nail or the stratum corneum (SC) electric properties since these three components of human tissues are dead keratinized substances, and they have similar electrical properties in frequencies above 50 kHz (Grimnes et al. 2011).

In the following sections, we will investigate the Electromagnetic (EM) properties of a three layer unit cell, with one cylindrical inclusion, which is irradiated with a plane wave normally incident on the skin. The spatial periodicity of the unit cell is considered in the (x-y) plane.

3 RESULTS

The unit cell is irradiated with a plane wave, where the wave vector k is parallel to the axis of the cylindrical inclusion z. Figure 2 depicts the magnitude of the E-field in three unit cells: the first one is without the cylindrical inclusion (Bold Unit Cell or BUC), the second is with a finite cylindrical inclusion (Shaved Unit Cell or SUC) and the last one is an infinite-cylindrical inclusion (Haired Unit Cell or HUC). The lengths of the hair shaft for SUC and HUC are respectively 500 um and 1.3 mm with a Mean Hair density (MHD) of 1135 per square centimeter and a cylinder radius of 40 um.

4 DISCUSSION

The E-field amplitude in Figure 2 is seen to be decreasing when both MHD and cylinder lengths are rising. A drop of more than −10 dB can be detected while varying the cylinder length by a factor of 20 times. SUC is showing higher electric amplitude compared to HUC, which may be summarized in the effect of the cylinder length and the distance between the cylinders (88 um) which is extremely small compared to the wavelength (10 mm), so for high values of MHD the distance from the skin to the head is the 4th layer of an effective material composed of hair and air which can be calculated using effective medium formulas (Markel 2016; Niklasson et al. 1981; Saviz et al. 2013).

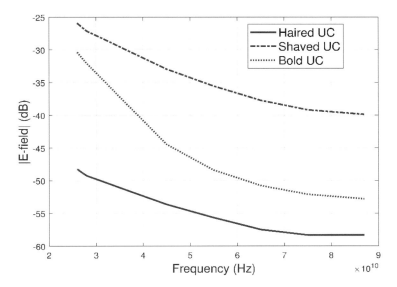

Figure 2.   E-Field evaluated in the interface between Fat and Muscle excited with a plane.

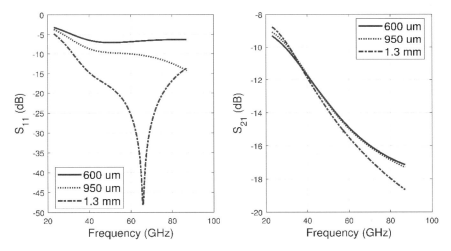

Figure 3.   S-parameters of the unit cell with a cylindrical inclusion density of 615 per square.

For BUC and SUC curves, we can prove using the Maxwell-Garnett formula as in equation 1, for a skin insertion in hair background material in which the resultant complex permittivity is lower than skin permittivity (see Figure 5). This fact results in lower E-Field values for SUC compared to the BUC.

$$\frac{\varepsilon_{eff} - \varepsilon_m}{\varepsilon_{eff} + 2\varepsilon_m} = \delta_i \frac{\varepsilon_i - \varepsilon_m}{\varepsilon_i + 2\varepsilon_m} \tag{1}$$

Where ε_{eff}, ε_m and ε_i are respectively the effective medium, the host medium and the insertion dielectric permittivity, δ_i is the inclusion factor.

Furthermore, the enhanced planar model can be seen from a Frequency Selective Surface perspective. The BUC model is showing notch-band capability in the K frequency bands with respect to the layer thicknesses of 500 um for skin and 200 um for both fat and muscle. Figure 4 shows the

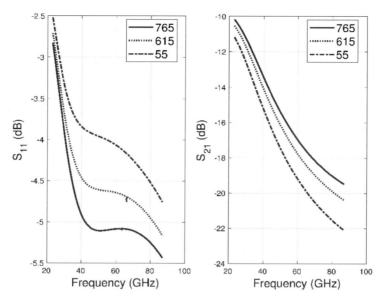

Figure 4. S-parameters of the unit cell with a cylindrical inclusion of different densities and a fixed length of 600 um.

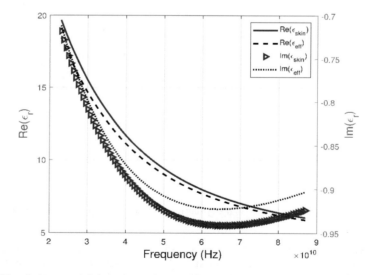

Figure 5. Skin relative permittivity compared to an effective medium composed of hair and skin. The cell dimension is 88 um and the mean hair density is 1135.

variation of S-parameters with density, while Figure 3 describes a variation with the length of the cylinder. An MHD variation is slightly effecting S plots, however the length of the cylinder plays much more of a part in decreasing S11 and rising S21. A pass-band capability can be obtained with a cylinder's length of 600 um.

5 CONCLUSION

The estimated E-field, using the model of Figure 1, is extremely dependent on the cylinder length and MHD. The 10 dB shift in E-field magnitude shows how in K-bands the human body model must embrace other components, which could have different dimensions and permittivity from hair follicles. However, most of the sub-components of the skin layer have either spherical (such as sebaceous glands) or cylindrical symmetries (such as hair). Furthermore, to properly use the effective medium theory, the cell extinction should be approximately zero.

REFERENCES

Andreuccetti, D., Fossi, R., & Petrucc, C. (1997). Dielectric Properties of Body Tissues: HTML clients. *IFAC-CNR, Florence, Italy*, 2007.

Blume-Peytavi, U., Whiting, D. A., & Trüeb, R. M. (Eds.). (2008). *Hair growth and disorders*. Springer Science & Business Media.

Grimnes, S., & Martinsen, O. G. (2011). *Bioimpedance and bioelectricity basics*. Academic Press.

Ho, H. S., Guy, A. W., Sigelmann, R. A., & Lehmann, J. F. (1971). Microwave heating of simulated human limbs by aperture sources. *IEEE Transactions on Microwave Theory and Techniques*, *19*(2), 224–231.

IEC Committee. IEEE 62704-1-2017-IEC/IEEE International Standard for Determining the Peak Spatial Average Specific Absorption Rate (SAR) in the Human Body from Wireless Communications Devices, 30 MHz-6 GHz.

IEEE Std C95. 3-2002. (2002). IEEE Recommended Practice for Measurements and Computations of Radio Frequency Electromagnetic Fields With Respect to Human Exposure to Such Fields, 100 kHz-300 GHz.

Joines, W. T., & Spiegel, R. J. (1974). Resonance absorption of microwaves by the human skull. *IEEE Transactions on Biomedical Engineering*, (1), 46–48.

Massoudi, H. (1979). Carlh. Durney, *"Electromagnetic Absorption in Multilayered Cylinderical Models of Man," IEEE Trans. Microwave Theory Tech*, *27*(10), 825–830.

Markel, V. A. (2016). Introduction to the Maxwell Garnett approximation: tutorial. *JOSA A*, *33*(7), 1244–1256.

Messaoudi, H., & Aguili, T. (2018, May). Use of a Split Ring Resonators with Dipole and PIFA Antenna to Reduce the SAR in a Spherical Multilayered Head Model. In *2018 6th International Conference on Multimedia Computing and Systems (ICMCS)* (pp. 1–6). IEEE.

Niklasson, G. A., Granqvist, C. G., & Hunderi, O. (1981). Effective medium models for the optical properties of inhomogeneous materials. *Applied Optics*, *20*(1), 26–30.

Saada, A. B., Mbarek, S. B., & Choubani, F. (2019, June). Antenna Polarization Impact On Electromagnetic Power Density for an Off-Body to In-Body Communication Scenario. In *2019 15th International Wireless Communications & Mobile Computing Conference (IWCMC)* (pp. 1430–1433). IEEE.

Sasaki, K., Wake, K., & Watanabe, S. (2014). Measurement of the dielectric properties of the epidermis and dermis at frequencies from 0.5 GHz to 110 GHz. *Physics in Medicine & Biology*, *59*(16), 4739.

Saviz, M., Toko, L. M., Spathmann, O., Streckert, J., Hansen, V., Clemens, M., & Faraji-Dana, R. (2013). A new open-source toolbox for estimating the electrical properties of biological tissues in the terahertz frequency band. *Journal of Infrared, Millimeter, and Terahertz Waves*, *34*(9), 529–538.

You, K. Y., & Then, Y. L. (2015). Electrostatic and dielectric measurements for hair building fibers from DC to microwave frequencies. *Int J Electr Comput Electron Commun Eng*, *9*, 337–344.

Innovative and Intelligent Technology-Based Services for Smart
Environments – Ben Slama et al (eds)
© 2021 Taylor & Francis Group, London, ISBN 978-1-032-02030-3

UWB textile antenna for medical applications

A. Mersani & W. Bouamra
MERLab LR18ES43, Department of Physics, University of Tunis El Manar, Tunisia

J.M. Ribero
LEAT, CNRS UMR 7248, Université Nice Côte d'Azur, Sophia Antipolis, France

L. Osman
MERLab LR18ES43, Department of Physics, University of Tunis El Manar, Tunisia

ABSTRACT: The purpose of this article is to propose and design a new low-profile antenna system composed of a miniature antenna associated with Artificial Magnetic Conductor (AMC) type compact metamaterial so as not to degrade antenna performance and reduce its sensitivity to its environment and to protect the human body against electromagnetic radiation from the antenna to the body. This structure has an Ultra Wide Band operation on the X band ranging from 8 GHz to 12 GHz and dedicated for medical applications.

1 INTRODUCTION

The field of telecommunications evolves more and more, and this requires a constantly increasing flow but is slowed down by an increasingly busy frequency spectrum (Li et al. 2018).

Recently, Ultra Wide Band technology has attracted great attention in both the industrial and medical fields. Indeed, UWB can reach speeds of several hundred megabits per second, while maintaining complexity and limited costs (Guo et al. 2018). Its impulse nature and bandwidth give it good resistance to interference and multiple paths, making it very suitable for indoor use. Also, its low spectral power density allows it to coexist by introducing little interference to surrounding systems. One of the key issues in UWB systems is to design appropriate antennas capable of operating in the desired frequency band. Several types of omnidirectional monopole antennas have been developed for short-distance communications. However, in the medical and military fields, and for detection applications, directional antennas have much more advantages. The improvement of communication and electronic technology has enabled the development of compact and intelligent antenna devices that can be positioned on the human body or implanted inside. A coplanar antenna with no reflector plane and back radiation must be isolated from the human body (Sievenpiper et al. 1999). Therefore, the coplanar antenna is always associated with an artificial magnetic conductor (AMC) structure (Mersani et al. 2017). The latter offers the potential advantage of improving radiation patterns, efficiency, bandwidth or even decreasing the size of the antenna (Alemaryeen & Noghanian 2017; Chen et al. 2015; Muhamad et al. 2018; Zhang et al. 2017).

This article focuses primarily on evaluating the performance of an antenna designed on the human body for medical applications. The goal is to study and design an antenna on an AMC artificial magnetic conductor to protect the human body. Table 1 illustrates acronyms listed in this paper.

2 ANTENNA DETAILS

A rectangular patch antenna design has been visualized. A felt-type substrate was chosen as the substrate of the antenna, having dimensions 32×36 mm2 with a dielectric constant $\varepsilon r = 1.22$ and

DOI 10.1201/9781003181545-10

Table 1.   Acronyms.

UWB	Ultra wide band
AMC	Artificial Magnetic Conductor
SAR	Specific absorption Rate

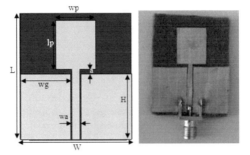

Figure 1.   Antenna prototype (W = 32, L = 36, lp = 14, wp = 11.5, wg = 14.8, wa = 2.22), all dimensions are in mm (Mersani et al. 2019).

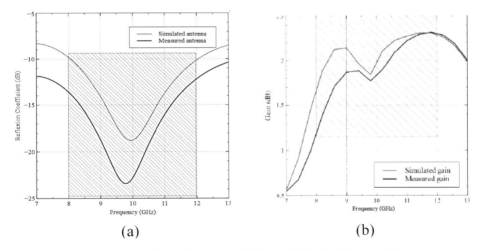

Figure 2.   Simulated and measured (a) reflection coefficient and (b) gain of the wearable antenna.

a thickness of H = 2 mm. Power is supplied via a coplanar line with an impedance normalized to 50Ω. The antenna has been optimized to resonate at ultra-wideband frequency from 8 GHz to 12 GHz.

The measurement of the reflection coefficient of the antennas was carried out with a vector network analyzer.

Figure 2(a) shows a comparison between the measurement results and the normalized reflection coefficient at 50 Ω of the antenna. There is good agreement between the measurement and the CST simulation of the reflection coefficient S11. The measured bandwidth −10 dB is between 7 GHz and 13 GHz, A slight shift can be observed at the level of the minimum of the adaptation which can be caused by the tolerance of manufacture and components. The measurements of the gain, efficiency and radiation patterns were made in the anechoic chamber of LEAT.

To calculate the gain of the antenna (GainANT), a comparison method is used using a standard antenna (reference antenna) whose value of realized gain is already known (GainREF).

$$GainANT = GainREF \times \frac{\text{Mesure ANT}}{\text{Mesure REF}} \qquad (1)$$

The maximum measured gain of textile antenna is 2.2 dBi as shown in Figure 2(b).

3 COMPACT AMC ANTENNA

3.1 *AMC structure*

A unit cell has been sized, is designed to have AMC-like behavior in the frequency band from 8 GHz to 12 GHz. It is a square patch and a circular slot in the center. The low cost felt substrate ($\varepsilon r = 1.22$, $\tan \delta = 0.0033$) was chosen for the design of each unit cell with a thickness of 2 mm for the felt and 0.0.35 mm for the lower substrate which is the Zelt (conductive part) as shown in Figure 3.

3.2 *AMC antenna*

In order to validate our structure, the proposed antenna a is placed at a distance of d = 4mm from a 3 × 4 cell AMC (as shown in Figure 4) and is realized and characterized. Its total dimensions are $36 \times 48 \times 8 \text{ mm}^3$.

Figure 5 illustrates a comparison between the measurement and the simulation of the reflection coefficient of the AMC.

Figure 6(a) shows a measured average realized gain equal to 5.6 dB (6 dB in simulation) throughout the X band. A slight shift in the maximum of the realized gain measured compared to the simulated one can also be noted. This small frequency shift is essentially due to the shift already identified at the level of the minimum of adaptation. Another important parameter for our study is the radiation efficiency. The total efficiency of the antenna as a function of frequency is calculated from the integration of antenna gain and directivity diagrams.

Figure 6(b) shows the evolution curves of measured and simulated radiation efficiency. The antenna has an average efficiency of 60%.

The dual question of the user impact on the antenna is that of the effect of the radiating element on the health of the person using the antenna. Although no study proves for the moment the risk generated by the use of devices with electromagnetic waves, international recommendations

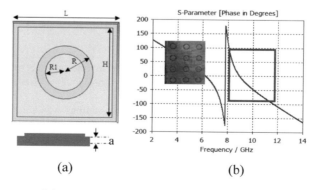

(a)   (b)

Figure 3.   (a) a prototype of the AMC unit cell and (b) the phase diagram. (L = 12mm, H = 11mm, R = 3.5mm, R1 = 3mm).

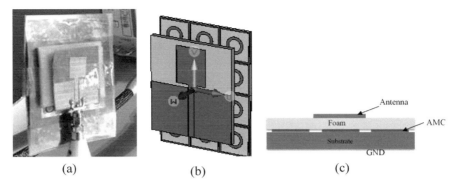

(a)   (b)   (c)

Figure 4. The proposed textile antenna, (a) the manufactured antenna, (b) the prototype of the monopole antenna, (c) the bottom view (Mersani et al. 2019).

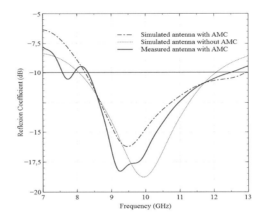

Figure 5. Reflection coefficient of the antenna with/without AMC.

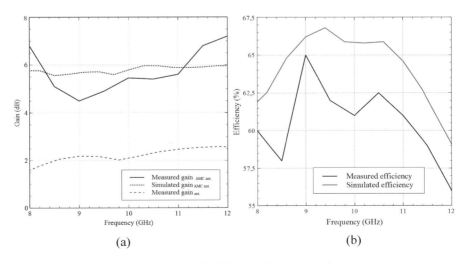

(a)   (b)

Figure 6. Simulated and measured: (a) gain, (b) efficiency of the monopole antenna.

Table 2. Result of max gain, directivity and SAR of the antenna with and without AMC.

	Antenna	Antenna + AMC
Gain (dB)	2.2	7
Directivity (dBi)	3.1	8.2
SAR w/kg (1 g tissue)	10.3	0.05

have identified the amount of electromagnetic power dissipated in the human body. For this, a measurement has been defined. This is the SAR (Specific Absorption Rate).

Table 1 summarizes the performance of the antenna with and without AMC. We note that with the addition of the the AMC structure, the performance of the antenna is improved. The gain of the antenna with AMC becomes 7 dB compared to 2.2 dB with the antenna alone. As well, we note that the SAR value has decreased from 10.3 w/kg to 0.05 W/Kg for 1g of tissue which proves that the body is well protected against the electromagnetic radiation of the antenna.

4 CONCLUSION

In this article, we relied on the first part on the study, the simulation and the manufacturing of an ultra wide band textile antenna. Then, in a logical sequence, a reflective plane of the artificial magnetic conductor type has been added to protect the human body against rear radiation and to improve the gain. To finally end up with a fairly complete prototype (antenna with AMC) which respects our initial specifications, namely: a low profile, low cost, lightness and ease of design on one side and which grants fairly interesting characteristics in terms adaptation of impedance, bandwidth, gain achieved, directivity and specific absorption rate (SAR) around the study frequency 5.8 GHz on the other. This antenna gives an acceptable performance to be integrated on the human body such as a good impedance adaptation, immunity against the proximity of the human body and a high gain going to 9 dBi.

REFERENCES

Alemaryeen, A. & Noghanian, S. 2017 "Performance analysis of textile AMC antenna on body model," 2017 USNC-URSI Radio Science Meeting (Joint with AP-S Symposium), San Diego, CA, pp. 41–42.

Chen, K., et al. 2015 "Improving microwave antenna gain and bandwidth with phase compensation metasurface", AIP Adv., 5, p. 067152.

Guo, L. Che, W & Yang, W 2018 "A Miniaturized Planar Ultra-Wideband Antenna," 2018 IEEE International Symposium on Antennas and Propagation & USNC/URSI National Radio Science Meeting, Boston, MA, pp. 1053–1054.

Li, Y. et al. 2018 "A compact triple wideband-notched UWB antenna," 2018 International Workshop on Antenna Technology (iWAT), Nanjing, pp. 1–4.

Mersani, A., Ribero, J.M & Osman, L. 2017 "Performance of Dual-Band AMC Antenna for WLAN Applications", IET Microwaves, Antennas & Propagation.

Mersani, A., Ribero, J.M & Osman, L., 2019 "Flexible UWB AMC Antenna for Early Stage Skin Cancer Identification", Progress In Electromagnetics Research M, Vol. 80, 71–81.

Muhamad, M., Abu, M & Zakaria, Z. 2018 "A Ground Plane of 5×5 Array AMC on a Single Antenna at 28 GHz," 2018 IEEE Student Conference on Research and Development (SCOReD), Selangor, Malaysia, pp. 1–5.

Sievenpiper, D. el al., 1999 "High-impedance electromagnetic surfaces with a forbidden frequency band," IEEE Transaction on Microwave Theory and Techniques, vol. 47, pp. 2059–2074, Nov.

Zhang, K. Zhou, X. Wei, Z. & Zhai, H. 2017 "A low-profile dual-band antenna loaded with the AMC surface," 2017 Sixth Asia-Pacific Conference on Antennas and Propagation (APCAP), Xi'an, pp. 1–3.

Innovative and Intelligent Technology-Based Services for Smart Environments – Ben Slama et al (eds)
© *2021 Taylor & Francis Group, London, ISBN 978-1-032-02030-3*

Wi-Fi-based human activity recognition using convolutional neural network

Muhammad Muaaz, Ali Chelli & Matthias Pätzold
Faculty of Engineering and Science, University of Agder, Grimstad, Norway

ABSTRACT: Unobtrusive human activity recognition plays an integral role in a lot of applications, such as active assisted living and health care for elderly and physically impaired people. Although existing Wi-Fi-based human activity recognition methods report good results, their performance is susceptible to changes in the environment. In this work, we present an approach to extract environment independent fingerprints of different human activities from the channel state information. First, we capture the channel state information by using the standard Wi-Fi network interface card. The channel state information is processed to reduce the noise and the impact of the phase offset. In addition, we apply the principal component analysis to removed redundant and correlated information. This step not only reduces the dimensions of the data but also removes the impact of the environment. Thereafter, we compute the spectrogram from the processed data which shows the environment independent fingerprint of the performed activity. We use these spectrogram images to train a convolutional neural network. Our approach is evaluated by using a human activity data set collected from 9 individuals while performing 4 activities (walking, falling, sitting, and picking up an object). The results show that our approach achieves an overall accuracy of 97.78%.

Keywords: Activity recognition; convolutional neural network; principal component analysis; spectrogram

1 INTRODUCTION

Wi-Fi-based human activity recognition (HAR) has become an important research topic due to the growing number of applications that need to monitor indoor human activities in a truly unobtrusive way. These applications include elderly care, surveillance, and active assisted living. Furthermore, Wi-Fi-based HAR offers several advantages over vision- and wearable sensor-based techniques. For instance, in contrast to vision-based HAR systems, Wi-Fi-based systems are cost-effective, unaffected by lighting conditions, and preserve the user's privacy. Furthermore, the users are not required to wear the sensor in contrast to wearable sensor-based HAR systems.

In Wi-Fi-based HAR systems, a transmitter and a receiver are deployed in the environment. The transmitter emits radio signals, and the presence of moving objects in the propagation environment causes the Doppler frequency shift in these radio signals before they are received by the receiver. In the literature, it has been shown that the received signal strength indicator (RSSI) and the channel state information (CSI) can be used to recognize human activities (Wang et al. 2017). In contrast to the RSSI which represents the attenuation of the received signal strength during propagation, the CSI is more informative. The CSI includes both amplitude and phase information associated with each orthogonal frequency division multiplexing (OFDM) subcarrier. It has been shown that CSI-based HAR systems generally perform better than RSSI-based HAR systems (Wang et al. 2017). There exist various approaches to recognize human activities from CSI data by using machine learning (Wang et al. 2017) and deep learning (Chen et al. 2018; Zou et al. 2018) techniques. In (Wang et al. 2017), the authors proposed two theoretical models. The first

DOI 10.1201/9781003181545-11

model (also known as "the CSI-speed model") links the speed of human body movements with the CSI data, while the second model (known as "the CSI-activity model") links the speed of human body movements with human activities (Wang et al. 2017). The proposed approach was developed using commercial Wi-Fi devices and achieved an overall recognition accuracy of 96%. In (Chen et al. 2018), an attention-based bidirectional long short-term memory (ABLSTM) technique was used to recognize humans activities from the CSI data. The CSI data sets collected in two different environments, namely an activity room and a meeting room were used to evaluate the performance of the proposed approach. This approach achieved recognition accuracies of 96.7% and 97.3% when the CSI data from the activity room and the meeting room is used, respectively. However, in the cross-environment scenario, where the training data that has been collected in one environment and the testing data from the other environment are used, the overall recognition accuracy drops to 32%. A deep learning technique consisting of autoencoder, convolutional neural network (CNN), and long short-term memory (LSTM) modules to recognize human activities from the CSI data has been proposed in (Zou et al. 2018). This deep learning network achieved an overall accuracy of 97.4%.

Although existing CSI-based HAR systems have reported reasonably good results, they still suffer from the drawback that they are environment dependent. This implies that their performance is susceptible to changes in the environment. One solution to this problem is to extract features from the CSI data that are subject and environment independent. This approach requires a lot of training data that must be collected from a variety of subjects in different environments (Jiang et al. 2018). The other approach proposes the use of a semi-supervised learning technique, which requires users to manually label the activity fingerprints that may have been changed due to changes in the environment (Wang et al. 2014). This solution requires user interaction that is not very practical for applications in elderly care.

In this work, we compute the spectrograms from the CSI data corresponding to different human activities. These spectrograms capture the Doppler characteristics of the radio channel caused by fixed and moving objects present in the environment. The static objects do not cause any variation in the trend of spectral components. This implies that different positions of static objects present in the environment will not influence the performance of our HAR system. These spectrograms are saved as portable network graphics (PNG) images and used to train a deep CNN. We evaluate this novel approach by using a CSI data set which is collected from 9 participants while performing four different activities: walking, falling, picking up an object, and sitting on a chair. Using this data set, our approach yields an overall accuracy of 97.78%. Moreover, our system recognizes the activities performed at greater distances. For instance, three out of the four activities are performed at a distance of 13 feet from the transmitting and receiving antennas.

The rest of the paper is organized as follows. Details about the experimental setup and human activity data collection are given in Section 2. In Section 3, we explain the steps involved in processing the CSI data and computing the spectrogram. In Section 4, we present our CNN model, the classification process, and the obtained results. Finally, concluding remarks are given in Section 5.

2 EXPERIMENTAL SETUP AND CSI DATA COLLECTION

In this paper, we considered an indoor environment where 9 participants performed four different activities, namely walking, sitting on a chair, falling on a mattress, and picking up an object from the floor. During the data collection process, we ensured that only a single person is moving inside the room and all other objects are static. The participants of this experiment were asked to stand still for one second before starting an activity and after finishing that activity.

For the walking activity, we asked the participant to walk in a straight line from Point A to Point B and back (see Figure 1). They repeated the activity 10 times, walking five times from Point A to B and five times from Point B to A. For the sitting activity, we placed a chair at Point B and asked the participants to stand still next to the chair facing the antennas and then sit on the chair as shown in Figure 1. For the falling activity, a mattress was placed at Point B, and the participants were asked

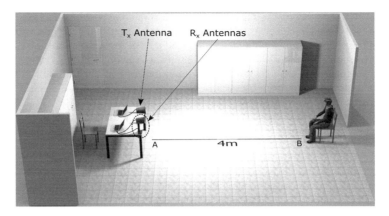

Figure 1. The experimental setup for data collection.

to stand on the shorter edge of the mattress and then fall on it. They repeated the activity 10 times. Out of these 10 falling trials, they fell on the mattress five times facing towards the antennas and five times facing away from the antennas. The last activity, picking up an object from the floor was also repeated five times, placing a small object on the floor at Point B, and asking the participants to pick it up.

To collect and parse the CSI data while the participants performed the activities mentioned above, we used two laptops, each was equipped with an Intel 5300 Wi-Fi network interface card (NIC). We installed the CSI Tool (Halperin et al. 2011) on both laptops. One laptop is used as the transmitter (T_x) and the other laptop as the receiver (R_x). The NICs of the T_x and the R_x are configured to operate at 5.745 GHz band with 20 MHz bandwidth in single-input multiple-output (SIMO) transmission mode. Instead of using the internal antennas of the laptops, which normally have a limited range, we connected an external directional antenna to the T_x and two external antennas to the R_x, where one of which was a directional and the other one an omnidirectional antenna. We used the injector-monitor Wi-Fi mode, where the T_x was set to inject 1000 random data packets per second into the wireless channel, and the R_x captured the injected packets. The transmitting and receiving antennas were attached to a table as shown in Figure 1 at a height of 0.8 meters from the floor. For each received data packet, the R_x reports the estimated CSI in a matrix form. The dimension of the CSI data matrix was $N_{T_x} \times N_{R_x} \times K$, where N_{T_x} indicates the number of transmit antennas, N_{R_x} stands for the number of receive antennas, and K represents the number of OFDM subcarriers. By default, the CSI tool reports estimated CSI data along 30 OFDM subcarriers for each transmission link. Therefore, in our case, the dimension of the CSI data matrix was $1 \times 2 \times 30$.

3 PROCESSING CSI DATA AND ESTIMATING THE SPECTROGRAM

The raw CSI data contains amplitude and phase information. Both the amplitude and phase of the CSI data are corrupted by noise; and therefore, the CSI data streams can not directly be used for activity recognition. The noise sources of the amplitude of the CSI data are mainly the ambient noise and adaptive changes of the transmission parameters (Yousefi et al. 2017). In addition to that, the phase of CSI data suffers from the carrier frequency offset (CFO) and the sampling frequency offset (SFO) (Wang et al. 2017; Yousefi et al. 2017). The errors related to the CFO and SFO are due to the asynchronicity between the transmitter and receiver clocks.

To denoise these data, we first calibrated the phase of the CSI data by applying the CSI ratio method (Zeng et al. 2019), where it has been shown that this approach significantly reduces the influence of CFO and SFO on the phase. The CSI ratio method requires that two antennas must be connected to the same receiver to collect the CSI data simultaneously. Thereafter, the CSI data from the first transmission link are divided by the CSI data of the second transmission link. Recall that each transmission link reports 30 CSI streams thus, we can obtain 30 CSI ratios. By comparing the spectrogram images of the CSI ratio method and the back-to-back phase calibration method (Keerativoranan et al. 2018), we observed in our experiments that the CSI ratio method works better.

Thereafter, we remove the correlated CSI ratios using the principal component analysis (PCA), which applies an orthogonal transformation to the 30 CSI ratios and converts them to 30 linearly uncorrelated variables. These variables are called principal components, where the first PCA component has the highest possible variance and the last PCA component the lowest variance. At this stage, we performed several experiments to determine the suitable number of PCA components for the subsequent steps. We observed that the first PCA component is sufficient to obtain a spectrogram that clearly shows the environment independent fingerprint of the performed activity as shown in Figure 2.

To further minimize the effect of the high frequency components, which are not caused by the human movement, we apply a low pass filter to the selected principal component. Thereafter, we first compute the short-time Fourier transform (STFT) of the filtered data as given in (1). In (1), t', t, $y(t)$, and $g(t)$ indicate the running time, the local time, the filtered data, and the Gaussian sliding window function, respectively.

$$X(f,t) = \int_{-\infty}^{\infty} y(t')g(t'-t)e^{-j2\pi ft'}dt'. \tag{1}$$

Finally, the STFT ($X(f,t)$) is multiplied with its complex conjugate ($S(f,t) = |X(f,t)|^2$) which gives the spectrogram (Boashash 2015).

4   CLASSIFYING SPECTROGRAM IMAGES WITH CNN

For every activity trial in the collected data, we first computed the spectrogram and then saved it as a PNG image in a folder labelled with the activity. Thereafter, all spectrogram images were scaled to the same $224 \times 224 \times 3$ dimension by applying the bicubic interpolation technique. We split the spectrogram data into the train, validation, and test data sets representing 70%, 15%, and 15% of the total data. The training data were used to train the CNN (shown in Figure 3) with a batch size of 16.

The CNN model consists of 14 layers including input, flatten, and output layers. The dimensions (i.e., height and width) of the filters used in all convolutional layers are 5×5 and in all max-pooling layers 2×2. The stride parameter (i.e., the number of cell shifts over the given data matrix) was set to 1 for the convolutional layers and to 2 for the max-pooling layers. The number of filters in the first, second, and third convolutional layer was 32, 48, and 64, respectively. All convolutional layers used the rectified linear unit (ReLU) activation function. After each max-pooling layer, a dropout layer (indicated by a green circle in Figure 3) with a threshold 0.3 was used. The last two layers are fully connected (FC) with dimensions 256×1 and 84×1, respectively. The dimension of the output layer is 4×1 and uses the softmax activation function. The validation data were used to monitor the training progress of the CNN and to stop the training if the validation accuracy does not improve over 8 consecutive epochs. The accuracy and loss of the CNN model over the training and validation data are presented in Figure 4(a) and Figure 4(b), respectively.

Finally, the performance of the CNN model was evaluated based on the test data set. The results of our approach are shown in the confusion matrix (see Table 1). In this confusion matrix, the green cells represent the correctly classified examples and incorrectly classified examples are indicated in the red cells. The overall accuracy of the CNN model is given in the blue diagonal cell. We observe

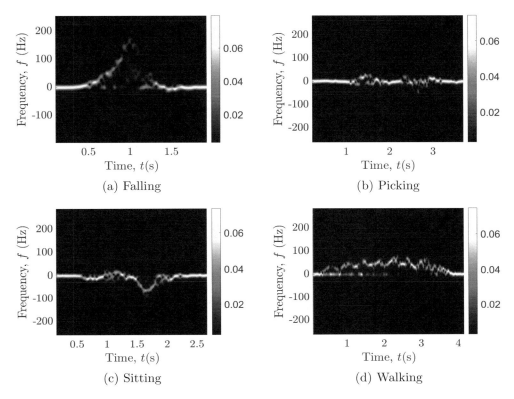

Figure 2. Spectrograms of the four activities.

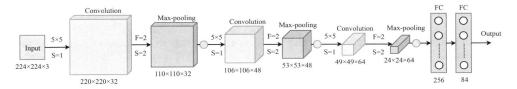

Figure 3. The architecture of the CNN, where the symbols S, F, and FC indicate stride, filter size of the max-pooling layer, and fully connected layer, respectively. The Green circles represent the dropout layer.

that the CNN model achieves an overall recognition accuracy of 97.78%. Moreover, the precision of the model for the activities walking, falling, picking up an object, and sitting is 100%, 93%, 100%, and 100%, respectively. The recall of the sitting activity is 88%, whereas the other three activities have a recall of 100%.

5 CONCLUSION

In this work, we developed a system that combines RF sensing and deep learning techniques to recognize human activities. In the RF sensing stage, we used two laptops, one acting as a transmitter and the other as a receiver to collect the CSI data. We collected CSI data while 9 participants performed 4 activities walking, falling, picking up an object from the ground, and sitting on a chair. A three-step process was used to filter the collected CSI data. At first, we applied the CSI

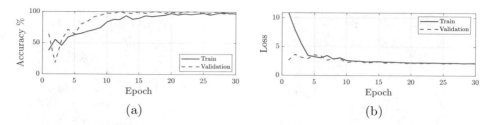

Figure 4. The accuracy (a) and loss (b) during the training process based on the training and validation data sets.

Table 1. The confusion matrix of results obtained from the CNN model.

-	-	Predicted labels				-
-	-	Walk	Fall	Pick	Sit	Precision
True labels	Walk	15	0	0	0	100%
	Fall	0	14	0	1	93%
	Pick	0	0	8	0	100%
	Sit	0	0	0	7	100%
-	Recall	100%	100%	100%	88%	97.78%

ratio method to the collected CSI data to reduce the impact of the phase offset. In the subsequent step, the PCA is applied to remove redundant and correlated information from the data. In the last step, a low pass filter is used to reduce the impact of high frequency components that were not caused by human movements. Thereafter, we computed a spectrogram for each activity trial in the collected data. These spectrogram images were divided into the train, validation, and test data sets. The training and validation data sets were used to train a 14-layer CNN model and monitor the training process, respectively. The test data set was used to evaluate the performance of the CNN model. The results show that our CNN model achieved an overall accuracy of 97.78%. In the future, we will conduct more experiments to quantitatively evaluate the performance of our approach in different environments.

ACKNOWLEDGEMENT

This work is carried out within the scope of WiCare project funded by the Research Council of Norway under the grant number 261895/F20.

REFERENCES

Boashash, Boualem. 2015. *Time-Frequency Signal Analysis and Processing – A Comprehensive Reference.* 2nd ed. Elsevier, Academic Press, Amsterdam.

Chen, Zhenghua, Le Zhang, Chaoyang Jiang, Zhiguang Cao, and Wei Cui. 2018. "WiFi CSI-Based Passive Human Activity Recognition Using Attention Based BLSTM." *IEEE Transactions on Mobile Computing* 18 (11): 2714–2724.

Halperin, Daniel, Hu Wenjun, Sheth Anmol, and Wetherall David. 2011. "Tool Release: Gathering 802.11N Traces with Channel State Information." *ACM SIGCOMM Comput. Commun. Rev.* 41 (1): 53–53.

Jiang, Wenjun, et al. 2018. "Towards Environment Independent Device Free Human Activity Recognition." In *Proc. of the 24th Int. Conf. on Mobile Computing and Networking*, 289–304.

Keerativoranan, Nopphon, Haniz Azril, Kentaro Saito, and Takada Junichi. 2018. "Mitigation of CSI Temporal Phase Rotation with B2B Calibration Method for Fine-Grained Motion Detection Analysis on Commodity Wi-Fi Devices." *Sensors* 18 (11).

Wang, Wei, X. Liu Alex, Shahzad Muhammad, Ling Kang, and Lu Sanglu. 2017. "Device-Free Human Activity Recognition Using Commercial WiFi Devices." *IEEE Journal on Selected Areas in Communications* 35 (5): 1118–1131.

Wang, Yan, et al. 2014. "E-eyes: Device-free Location-oriented Activity Identification Using Fine-grained WiFi Signatures." In *Proc. of the 20th Int. Conf. on Mobile Computing and Networking*, 617–628.

Yousefi, Siamak, Narui Hirokazu, Dayal Sankalp, Ermon Stefano, and Valaee Shahrokh. 2017. "A Survey on Behavior Recognition Using WiFi Channel State Information." *IEEE Communications Magazine* 55 (10): 98–104.

Zeng, Youwei, et al. 2019. "FarSense: Pushing the Range Limit of WiFi-Based Respiration Sensing with CSI Ratio of Two Antennas." *In Proc. ACM Interact. Mob. Wearable Ubiquitous Technol.* 3 (3).

Zou, Han, et al. 2018. "Deepsense: Device-free Human Activity Recognition via Autoencoder Long-term Recurrent Convolutional Network." In *2018 IEEE International Conference on Communications (ICC)*, 1–6.

Innovative and Intelligent Technology-Based Services for Smart Environments – Ben Slama et al (eds)
© 2021 Taylor & Francis Group, London, ISBN 978-1-032-02030-3

A new non-stationary 3D channel model with time-variant path gains for indoor human activity recognition

R. Hicheri, A. Abdelgawwad & M. Pätzold
Faculty of Engineering and Science, University of Agder, Grimstad, Norway

ABSTRACT: This paper introduces a new three-dimensional channel model to describe the propagation phenomenon taking place in indoor environments, equipped with a multiple-input multiple-output system, in the presence of a single moving person. Specifically, assuming that the person is modelled by a cluster of moving point scatterers, which act as relays, we derive the time-variant transfer function with time-variant path gains, time-variant Doppler shifts, and time-variant path delays. Then, we analyse the effect of the motion of the person on the spectrogram and the time-variant mean Doppler shift. Simulation results are presented to illustrate the proposed channel model.

Keywords: Indoor propagation, non-stationary channels, cluster of moving scatterers, human activity recognition.

1 INTRODUCTION

With the expectation that the elderly population will triple in the next 30 years [United Nations (Oct. 2015)], there has been great interest in the study of home care systems based on human activity recognition, tracking, and classification techniques. A review of the literature shows that human activity recognition systems utilize two major techniques: wearable inertial measurement units and video surveillance [Jobanputra, Bavishi, and Doshi (2019)]. The main drawbacks of these systems are: the person may forget to wear the sensors-based devices and/or go outside the cameras' coverage area [Seneviratne et al. (4th Quart., 2017)]. Non-wearable radio-frequency (RF)-based systems have been introduced to overcome the drawbacks of the aforementioned systems [Liu, Liu, Chen, Wang, and Wang (Aug. 2019)]. These RF-based devices track and monitor indoor human activities by exploiting the effects of the human motion on the Doppler characteristics of the transmitted RF signals.

A three-dimensional (3D) non-stationary channel model with constant path gains has been reported in [Abdelgawwad and Pätzold (Apr. 2019)], where the body segments of the moving person have been modelled by a cluster of synchronized moving scatterers. The authors analysed the effects of the motion of the person on the spectrogram of the complex channel gain. A 3D non-stationary channel model with a stochastic trajectory model has been presented in [Borhani and Pätzold (Nov. 2018)] which incorporates time-variant (TV) path gains that change with the distance of the moving person from the transmitter/receiver. In [Borhani and Pätzold (Nov. 2018)], the TV path gains are determined according to the path loss model in [Phillips, Sicker, and Grunwald (1st Quart., 2013)] by considering the sum of the travelled distances between the scatterers and the transmitter/receiver. This assumption does not match the results of several studies, in which the effects of electromagnetic fields on the human body have been investigated investigated [Council et al. (1993a, 1993b); Federal Public Service (Jan. 2019); Mahmoudinasab, Sanie-Jahromi, and Saadat (Jun. 2016)]. According to the aforementioned references, in the presence of low-frequency electromagnetic fields, an electric current is generated in the human body which is generally referred to as "induced current". As a result, the different body segments of the moving person are assumed

DOI 10.1201/9781003181545-12

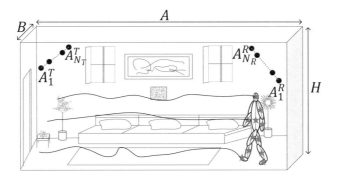

Figure 1. A single moving person (modelled by a cluster of moving scatterers) walks in a room equipped with a MIMO system and several fixed (stationary) objects.

to play the role of relays, which redirect the transmitted RF signal. In this case, the mean power of the TV path gain of each multipath component is modelled according to the free-space path loss model [Phillips et al. (1st Quart., 2013)], which depends on the product of the distance between the transmit antenna and the scatterer (body segment) and the distance between the scatterer and the receive antenna.

The goal of this work to address this problem. To do so, we present a new generic non-stationary 3D indoor channel model, where the different body segments of the moving person are assumed to play the role of moving relays. In this case, we present an expression for the time-variant transfer function (TVTF) with TV path gains and TV path delays, where the mean power of TV path gains is modelled by the free-space path loss model [Phillips et al. (1st Quart., 2013)].

The analysis of the influence of the motion of different body segments on the Doppler characteristics of the channel is conducted by means the spectrogram of the TVTF and the TV mean Doppler shift. Simulation results are presented to illustrate the proposed channel model.

The remaining of this paper is organised as follows. Section II discusses the indoor propagation scenario with the cluster of moving scatterers. The derivation of the TVTF of the received RF signals is presented in Section III. Section IV presents some numerical results and Section V concludes the paper.

2 SCENARIO DESCRIPTION

As illustrated in Figure 1, we consider a rectangular cuboid room. The room of length A, width B, and height H is equipped with an $N_T \times N_R$ MIMO communication system, where N_T denotes the number of transmit antennas A_j^T ($j = 1, \ldots, N_T$) and N_R designates the number of receive antennas A_i^R ($i = 1, \ldots, N_R$). There are several stationary (fixed) objects (e.g., walls, furniture, and decoration items) and a single moving person. The moving person is modelled by a cluster of moving scatterers, where each single point scatterer (★) represents a segment of the human body such as head, shoulders, arms, etc.

The corresponding geometrical 3D channel model shown in Figure 2 is the starting point for modelling the indoor propagation phenomenon. Each transmit (receive) antenna element A_j^T (A_i^R) is located at the coordinates (x_j^T, y_j^T, z_j^T) ((x_i^R, y_i^R, z_i^R)), $j = 1, 2, \ldots, N_T$ ($i = 1, 2, \ldots, N_R$). Here, we assume the presence of LOS components. The fixed objects between A_i^T and A_j^R are modelled by K_{ij} fixed point scatterers (•), denoted by $S_{k_{ij}}^F$, $k_{ij} = 1, 2, \ldots, K_{ij}$. On the other hand, every body segment of the moving person is modelled by a single point moving scatterer (★), denoted by S_n^M ($n = 1, 2, \ldots, M$), which is located at the TV position $(x_n(t), y_n(t), z_n(t))$. Single-bounce scattering is assumed when modelling the fixed and moving scatterers.

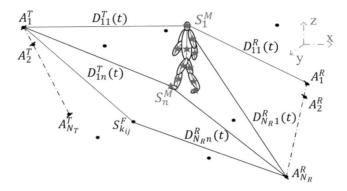

Figure 2. The 3D geometrical model for an $N_T \times N_R$ MIMO channel with fixed ($\bullet$) and moving scatterers ($\star$).

Together with considering TV path gains and TV path delays, the other novelty of this work is that the human body segments are assumed to play the role of relays, which redirect the transmitted radio wave. This is motivated by the fact that low-frequency electromagnetic fields have been shown to induce (in)perceivable electric currents in the human body, which in turn can cause a tingling sensation depending on the frequency range [Council et al. (1993a, 1993b); Federal Public Service (Jan. 2019); Mahmoudinasab et al. (Jun. 2016)]. In other words, when the transmitted RF signal impinges on a body segment (moving scatterer), then an electric current induced, which retransmits the RF signal in all directions. In the presence of these induced currents, the human body segments (moving scatterers) play the role of relays. In this case, the free-space path loss model [Phillips et al. (1st Quart., 2013)] states that the power $P_{in}^R(t)$ received at the ith receive antenna A_i^R after being relayed by the nth moving scatterer S_n^M is given by

$$P_{in}^R(t) = a_n^2 C^2 \left[D_{in}^R(t) \times D_{jn}^T(t) \right]^{-\gamma} \tag{1}$$

where the quantity C depends on the transmit/receive antenna gain, the transmission power, and the wavelength [Phillips et al. (1st Quart., 2013)] and γ is the path loss exponent which is equal to 2 in free space and between 1.6 a1.8 in indoor environments. The parameter a_n describes the weight (contribution) of the nth moving scatterer S_n^M with respect to the total propagation phenomenon, with $\sum_{n=1}^M a_n^2 = 1$. Here, the quantity $D_{jn}^T(t)$ ($D_{in}^R(t)$) represents the TV distance between the jth (ith) transmit (receive) antenna A_j^T (A_i^R) and the nth moving scatterer S_n^M. According to Figure 2, the TV distances $D_{jn}^T(t)$ and $D_{in}^R(t)$ can be expressed as

$$D_{jn}^T(t) = \left[\left(x_n(t) - x_j^T \right)^2 + \left(y_n(t) - y_j^T \right)^2 + \left(z_n(t) - z_j^T \right)^2 \right]^{1/2} \tag{2}$$

and

$$D_{in}^R(t) = \left[\left(x_i^R - x_n(t) \right)^2 + \left(y_i^R - y_n(t) \right)^2 + \left(z_i^R - z_n(t) \right)^2 \right]^{1/2} \tag{3}$$

respectively.

3 THE 3D NON-STATIONARY CHANNEL MODEL

According to Figure 2, the TVTF $H_{ij}(f', t)$ of the sub-channel between A_j^T and A_i^R can be expressed as

$$H_{ij}(f', t) = \sum_{n=1}^{M} c_{ijn}(t) \exp\left(j\left(\theta_{ijn} - 2\pi f' \tau'_{ijn}(t)\right)\right) + \sum_{k=0}^{K_{ij}} c_{k_{ij}} \exp\left(j\left(\theta_{k_{ij}} - 2\pi f' \tau'_{k_{ij}}\right)\right). \tag{4}$$

The first part of (4) describes the effects of the motion of the cluster of scatterers (representing the moving person). Here, the quantities $c_{ijn}(t)$ and $\tau'_{ijn}(t)$ are the TV path gain and TV path delay of the nth moving scatterer S_n^M ($n = 1, 2, \ldots, M$), respectively. These parameters are given by

$$c_{ij}(t) = \sqrt{P_{in}^R(t)} = a_n C \left[D_{in}^R(t) \times D_{jn}^T(t)\right]^{-\gamma/2} \tag{5}$$

and

$$\tau'_{ijn}(t) = \frac{D_{jn}^T(t) + D_{in}^R(t)}{c_0}. \tag{6}$$

The quantity c_0 denotes the speed of light. The initial phases of the channel, denoted by θ_{ijn} are modelled by independent and uniformly distributed (i.i.d.) random variables which are uniformly distributed over the interval $[0, 2\pi)$.

The second sum of (4) models the multipath propagation effect resulting from the fixed scatterers. Each fixed scatterer $S_{k_{ij}}$ is described by a constant path gain $c_{k_{ij}}$, a constant path delay $\tau'_{k_{ij}}$, and a constant phase $\theta_{k_{ij}}$, $k = 1, \ldots, K_{ij}$. The LOS component, which doesn't experience any Doppler effect, can be modelled by a fixed scatterer S_0^F ($k = 0$), with a constant path gain $c_{0_{ij}}$ and a constant delay $\tau_{0_{ij}}$. The phases $\theta_{k_{ij}}$ are modelled by i.i.d. random variables with a uniform distribution over the interval $[0, 2\pi)$. For simplicity, the overall effect of the fixed scatterers, i.e., the second sum of (4), is replaced by a single complex term $c_{ijF}(f') \exp\left[j(\vartheta_{ijF}(f'))\right]$, with an envelope $c_{ijF}(f')$ and a phase $\vartheta_{ijF}(f')$. These quantities are expressed as $c_{ijF}(f') = \{[\sum_{k=0}^{K_{ij}} c_{k_{ij}} \cos(\theta_{k_{ij}} - 2\pi f' \tau'_{k_{ij}})]^2 + [\sum_{k=0}^{K_{ij}} c_{k_{ij}} \sin(\theta_{k_{ij}} - 2\pi f' \tau'_{k_{ij}})]^2\}^{1/2}$ and $\vartheta_{ijF}(f') = \text{atan2}(\sum_{k=0}^{K_{ij}} c_{k_{ij}} \sin(\theta_{k_{ij}} - 2\pi f' \tau'_{k_{ij}}), \sum_{k=0}^{K_{ij}} c_{k_{ij}} \cos(\theta_{k_{ij}} - 2\pi f' \tau'_{k_{ij}}))$, respectively, where atan2$) \cdot)$ is the inverse tangent function.

To visualize the influence of the moving person (cluster of scatterers), we consider the spectrogram $S_{ij}(f, t)$ of the TVTF $H_{ij}(f', t)$ using a Gaussian window [Abdelgawwad and Pätzold (Apr. 2019); Borhani and Pätzold (Nov. 2018)]. For short time intervals during which the coordinates $x_n(t)$, $y_n(t)$, and $z_n(t)$ of the moving scatterers S_n^M ($n = 1, 2, \ldots, M$) can be assumed to be constant, an approximate solution to the spectrogram $S_{ij}(f, t)$ of the TVTF $H_{ij}(f', t)$ can determined according to [(Borhani & Pätzold, Nov. 2018, Equations (17)(20))]. From [Borhani and Pätzold (Nov. 2018)], the TV mean Doppler shift $B_{ij,H}^{(1)}(t)$ can be determined from the spectrogram $S_{ij}(f, t)$ using [(Borhani & Pätzold, Nov. 2018, Equation (22))].

4 SIMULATION RESULTS

In this section, we explore the propagation scenario described in Section 2 by studying the effect of the TV path gains on the spectrogram and the TV mean Doppler shift of the TVTF of the received RF signal. For brevity, the transmitter T_X and the receiver R_X are co-located on the ceiling of the room and equipped with a single antenna. A single walking person is considered in two different scenarios. In the first one, the person is represented by a single moving scatterer ($M = 1$) corresponding to the head. In the second one, the person is modelled by 6 moving point scatterers

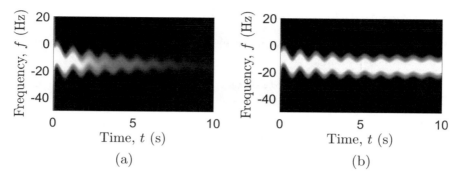

Figure 3. Spectrogram $S_{11}(f,t)$ for $M = 1$ with (a) TV path gains and (b) constant path gains.

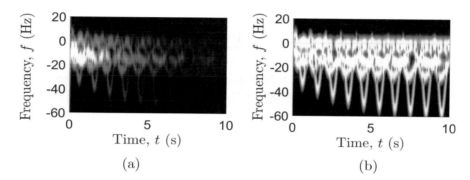

Figure 4. Spectrogram $S_{11}(f,t)$ for $M = 6$ with (a) TV path gains and (b) constant path gains.

($M = 6$) representing the major segments of the human body, namely, the head, the ankles, the wrists, and the waist. The trajectories of the different body segments are simulated according to [(Abdelgawwad & Pätzold, Apr. 2019, Section V)]. The person is initially located below to the T_X and R_X antennas and walks away from the antennas. The parameters C and γ are set to 1 and 1.6, respectively. The contribution of the fixed scatterers is removed by using a high-pass filter. For the remaining parameters as well as the scenario with constant path gains, we consider the same simulation parameters as in [(Abdelgawwad & Pätzold, Apr. 2019, Section V)].

Figures 3(a) and (b) (Figures 4(a) and (b)) depict the spectrogram $S_{11}(f,t)$ corresponding to Scenario I (Scenario II) while considering TV path gains and constant path gains, respectively. In Figures 3 and 4, the Doppler frequencies caused by the walking activity have negative values because the propagation delays (total travelled distances) increase when the person is walking away from the T_X and R_X antennas. By comparing Figures 3(a) and 4(a) (Figures 3(b) and 4(b)), it can be observed that the resolution of the spectrogram is lower for Scenario II ($M = 6$). This is due to the presence of the spectrogram's cross-term (see [Abdelgawwad and Pätzold (Apr. 2019), Equation (47)]). From Figures 3(a) and 4(a), it can also be noted that the instantaneous power of the spectrogram decreases w.r.t. time t. This stems from the fact that when the person is moving away from the antennas, the TV distances $D_{1n}^T(t)$ and $D_{1n}^R(t)$ increase w.r.t. time t. In practice, this can be explained by the fact that by moving away from the T_X and the R_X antennas, the person is walking out of the range of the RF system. This is in contrast to what can be observed in Figures 3(b) and 4(b), where the instantaneous power of the spectrogram remains constant, i.e., the range of the RF system is infinite.

To further investigate the effect of introducing TV path gains on the Doppler characteristics of channel, we present in Figures 5(a) and (b) a comparison of the TV mean Doppler shifts $B_{11,H}^{(1)}(t)$

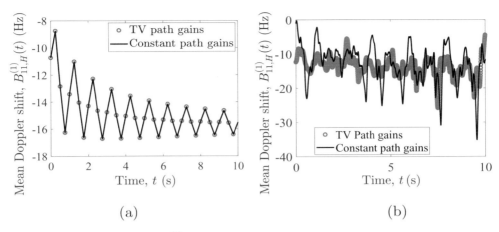

Figure 5.   TV mean Doppler shift $B_{11,H}^{(1)}(t)$ for (a) $M = 1$ and (b) $M = 6$.

using TV path gains and constant path gains for Scenario I ($M = 1$) and Scenario II ($M = 6$), respectively. From Figure 5(a), it is clear that for $M = 1$, the TV behavior of the path gains does not affect $B_{11,H}^{(1)}(t)$. This comes from the fact that for $M = 1$, $B_{11,H}^{(1)}(t)$ reduces to the TV Doppler frequency $f_{11}(t)$ (see [Borhani and Pätzold (Nov. 2018), Equation (21)]). The effect of the TV path gains can be clearly seen in Figure 5(b) for $M = 6$. In fact, for $M > 1$, the TV path gains affect both the trend and the values of $B_{11,H}^{(1)}(t)$.

ACKNOWLEDGEMENT

This work was supported by the WiCare Project funded by the Research Council of Norway under grant number 261895/F20.

REFERENCES

Abdelgawwad, A., & Pätzold, M. (Apr. 2019). A 3D non-stationary cluster channel model for human activity recognition. In *IEEE 89th Veh. Technol. Conf. (VTC'19-Spring)* (pp. 1–7).

Borhani, A., & Pätzold, M. (Nov. 2018). A non-stationary channel model for the development of non-wearable radio fall detection systems. *IEEE Trans. Wireless Commun., 17*(11), 7718–7730.

Council, N. R., et al. (1993a). Effects of electromagnetic fields on organs and tissues. In *Assessment of the possible health effects of ground wave emergency network*. National Academies Press (US).

Council, N. R., et al. (1993b). Human laboratory and clinical evidence of effects of electromagnetic fields. In *Assessment of the possible health effects of ground wave emergency network*. National Academies Press (US).

Federal Public Service. (Jan. 2019). *Federal Public Service: Interaction between radiation and the human body*. (Accessed in: Mar. 2020. Available: https://www.health.belgium.be/en/interaction-between-radiation-and-human-bodyarticle)

Jobanputra, C., Bavishi, J., & Doshi, N. (2019). Human activity recognition: A survey. *Procedia Computer Science, 155*, 698–703.

Liu, J., Liu, H., Chen, Y., Wang, Y., & Wang, C. (Aug. 2019). Wireless sensing for human activity: A survey. *IEEE Commun. Surveys Tuts*.

Mahmoudinasab, H., Sanie-Jahromi, F., & Saadat, M. (Jun. 2016). Effects of extremely lowfrequency electromagnetic field on expression levels of some antioxidant genes in human MCF-7 cells. *Molecular Biology Research Commun., 5*(2), 77–85.

Phillips, C., Sicker, D., & Grunwald, D. (1st Quart., 2013). A survey of wireless path loss prediction and coverage mapping methods. *IEEE Commun. Surveys & Tuts.*, *15*(1), 255–270.

Seneviratne, S., et al. (4th Quart., 2017). A survey of wearable devices and challenges. *IEEE Commun. Surveys & Tuts.*, *19*(4), 2573–2620.

United Nations. (Oct. 2015). *United Nations: Transforming our world: the 2030 agenda for sustainable development*. (Accessed in: Mar. 2020. [Online]. Available: https://www.health.belgium.be/en/interaction-between-radiation-and-humanbody# article, report no.A/RES/70/1).

Innovative and Intelligent Technology-Based Services for Smart Environments – Ben Slama et al (eds)
© 2021 Taylor & Francis Group, London, ISBN 978-1-032-02030-3

Human activity recognition using Wi-Fi and machine learning

Ali Chelli, Muhammad Muaaz, Ahmed Abdelgawwad & Matthias Pätzold
Faculty of Engineering and Science, University of Agder, Grimstad, Norway

ABSTRACT: In the era of Internet of things, access points will be deployed everywhere. The wireless signals offered by these access points can be used for more than just Internet connectivity. In fact, the human movement causes Doppler shifts in the received wireless signals. By combining signal processing techniques and machine learning, it is possible to recognize human activity from Wi-Fi signals. This paper builds on these ideas and develops a human activity recognition system that comprises two parts: radio-frequency sensing and machine learning. In the radio-frequency sensing part, we record the channel transfer function of an indoor environment in the presence of a participant performing three activities: walking, falling, and picking up an object. Using signal processing techniques, we estimate the mean Doppler shift of the channel, which contains the fingerprint of the user activity. The mean Doppler shift is used by a classifier to determine the type of performed activity. We assess the activity recognition performance of three classification algorithms: cubic support vector machine, K-nearest neighbor, and linear discriminant analysis. Our analysis shows that the cubic support vector machine, linear discriminant analysis, and K-nearest neighbor algorithms achieve an overall accuracy of 99.5%, 97.3%, and 95.1%, respectively.

Keywords: Human activity recognition, machine learning, channel state information, channel transfer function, mean Doppler shift.

1 INTRODUCTION

With the rapid development of the Internet of things (IoT), various types of sensors have been embedded in indoor and outdoor environments. This offers the opportunity to collect useful data that can be utilized in environment monitoring, smart city, and human activity recognition (HAR). Traditional HAR techniques are mainly based on camera (Sehairi, Chouireb, and Meunier 2017) or wearable devices (Chelli and Pätzold 2019). However, with the widespread deployment of Wi-Fi access points, device-free HAR has attracted a lot of attention. As opposed to camera-based HAR, Wi-Fi-based HAR does not violate user's privacy. Besides, Wi-Fi-based HAR does not require the user to wear any sensing device, which allows avoiding user discomfort associated with wearable devices.

In Wi-Fi-based activity recognition systems, an electromagnetic wave emitted by the transmitter is reflected by the objects in the environment before reaching the receiver. If a person is moving in the vicinity of the transmitter and the receiver, this movement causes a Doppler shift in the received radio-frequency (RF) signal. The pattern of this Doppler shift varies depending on the type of activity. Thus, it is possible to recognize the user activity based on the received RF signal. Wi-Fi-based HAR either uses the channel state information (CSI) (Wang, Gong, and Liu 2019; Zou et al. 2018; Zou et al. 2018) or the radio signal strength indicator (RSSI) (Abdelnasser, Youssef, and Harras 2015) for activity recognition. CSI-based HAR systems have a better accuracy in recognizing human activity compared to RRSI-based HAR systems. The authors of (Halperin et al. 2011) developed a software tool for gathering CSI from Wi-Fi data packets by means of a network interface card (NIC). The authors of (Wang, Gong, and Liu 2019), (Zou et al. 2018), and

(Zou et al. 2018) utilize deep learning algorithms for activity recognition based on CSI data and achieve an overall recognition accuracy of 96%, 97.4%, and 97.6%, respectively.

In this paper, we use the software tool proposed in (Halperin et al. 2011) to build a testbed for HAR using two laptops. One laptop acts as a transmitter, while the second laptop acts as a receiver. Seven participants are asked to perform three activities (walking, falling, and picking up an object) inside a 16 square meter room. The CSI data is collected while the participants carry out their activities. Using signal processing techniques, we estimate the time-variant mean Doppler shift (MDS) from the recorded CSI data. Subsequently, time and frequency domain features are extracted from the MDS and provided to a classification algorithm. This classifier must determine the type of performed activity. We test the performance of three classification algorithms: cubic support vector machine (CSVM), K-nearest neighbor (KNN), and linear discriminant analysis (LDA). Our results show that the CSVM, LDA, and KNN algorithms achieve an overall recognition accuracy of 99.5%, 97.3%, and 95.1%, respectively. With these results, we outperform most of the existing activity recognition systems in terms of overall accuracy.

Note that this is the first work in the literature that uses the MDS for activity recognition. The MDS approach makes our system robust to changes in the environment. In other words, if the position of the fixed scatterers (e.g., furniture, walls) is modified, this has no impact on the recognition accuracy. Moreover, our investigation reveals that our HAR system can maintain a high recognition accuracy even if the activity is carried out at a distance of 4 meters from the transmitter and receiver. Note that in most existing studies, the activity is performed very close (one meter) to the transmitter and receiver to achieve a recognition accuracy of over 90%. It is reported in (Wang, Wu, and Ni 2017) that the recognition accuracy drops below 85% if the participant is at a distance of 3 meters from the transmitter and receiver.

The rest of the paper is organized as follows. Section 2 provides an overview of the proposed activity recognition system and its building components. In Section 3, we describe the different data pre-processing steps that are applied to the collected CSI data to obtain the MDS. In Section 4, we assess the performance of the proposed HAR system and discuss the obtained results. Finally, Section 5 offers concluding remarks.

2 OVERVIEW OF THE CSI-BASED HAR SYSTEM

An overview of the CSI-based HAR system is provided in Figure 1. The HAR system uses the CSI data of a Wi-Fi link to recognize human activities. The HAR system consists of two main stages: RF sensing and machine learning as shown in Figure 1. In the RF sensing stage, a single person carries out an activity in an indoor environment. The RF data is collected by involving seven participants performing three different activities: walking, falling, and picking up an object. During the activity, the person's body parts move and cause a Doppler shift in the RF signal. The pattern of this Doppler shift varies from one activity to another and can thus be used for activity recognition. To capture this phenomenon, we use two laptops acting as a transmitter and receiver. Instead of the built-in RF antennas of the laptops, we use RF cables and connected external omni-directional antennas to the NICs of the laptops. A single transmit antenna A_T injects 1000 data packets per second in the wireless medium and two receive antennas A_{R1} and A_{R2} collect the injected packets. All the antennas are attached to the room ceiling. We use the software tool proposed in (Halperin et al. 2011) to capture the Wi-Fi CSI from the received packets, while the users perform different activities.

The CSI data recorded during RF sensing is then transferred to the machine learning part. Using signal processing algorithms, the CSI data is calibrated, filtered, denoised, and the MDS associated with each CSI sample is computed. The MDS contains the fingerprints of the user activity and its pattern varies as the activity changes. The MDS is considered as a time-series from which we extract time and frequency domain features that form the feature vector. Based on this vector, the classification algorithm must determine the type of performed activity from three possible activities: walking, falling, and picking up an object (see Figure 1).

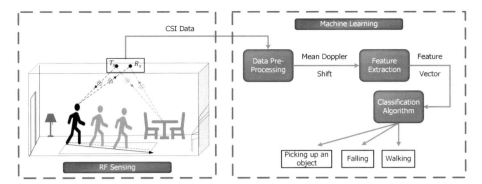

Figure 1.   Architecture of the HAR system.

3   DATA PRE-PROCESSING: FROM CSI DATA TO MDS

The transmitter laptop T_x sends 1000 data packets per second in the wireless medium. These packets are transformed into an electromagnetic wave by the transmit antenna A_T. This wave travels in the indoor environment and is reflected by fixed (walls and furniture) and moving (body parts of the moving person) objects before arriving at the receive antennas A_{R1} and A_{R2} of the receiver laptop R_x. To capture the CSI of the received data packets, we use the software tool developed in (Halperin et al. 2011). The CSI data is an $N_{T_x} \times N_{R_x} \times K$ matrix, where N_{T_x} is the number of transmit antennas, N_{R_x} denotes the number of receive antennas, and K stands for the number of orthogonal frequency division multiplexing (OFDM) subcarriers.

In our measurement campaign, we have one transmit antenna ($N_{T_x} = 1$), two receive antennas ($N_{R_x} = 2$), and 30 OFDM subcarriers ($K = 30$). The symbol f'_k refers to the carrier frequency of the kth OFDM subcarrier. We denote the elements of the CSI matrix by $H_{i,j}(f'_k, t)$, where the pair (i, j) indicates the indices of the transmit and receive antenna, respectively. Each element $H_{i,j}(f'_k, t)$ of the CSI matrix is known as a CSI stream. The link between the ith transmit antenna and the jth receive antenna is characterized by its channel transfer function (CTF) $H_{i,j}(f', t)$. Thus, the CSI stream $H_{i,j}(f'_k, t)$ is a discrete version of the CTF $H_{i,j}(f', t)$ sampled at frequency f'_k.

In our measurement campaign, the distance between the moving person and the antennas is short (less than 4 meters), which makes the angles of departure and arrival time variant. Besides, the speed of motion of different body parts is also time variant. This makes the underlying fading channel non-stationary, which implies that both the CTF and the CSI stream are non-stationary. Therefore, a time-frequency analysis tool such as the spectrogram is needed to analyse the behavior of the fading and characterize the fingerprint related to various user activities.

The spectrogram $S_{H_k}(f, t)$ of the CSI stream $H_{i,j}(f'_k, t)$ is obtained in two steps. First, we compute the short-time Fourier transform (STFT) $X_{H_k}(f, t)$ of the CSI stream $H_{i,j}(f'_k, t)$

$$X_{H_k}(f, t) = \int\limits_{-\infty}^{\infty} H_{i,j}(f'_k, \tau) w(\tau - t) e^{-j2\pi f \tau} d\tau \qquad (1)$$

where τ stands for the running time, t is the local time, and $w(t)$ is a sliding window. In our case, we use a Gaussian window function (Boashash 2015, Equation (2.3.1)). Second, the spectrogram

Table 1. Performance of the classification algorithms KNN, LDA, and CSVM in terms of activity recognition.

		Algorithms		
		KNN	LDA	CSVM
Overall Accuracy %		95.1	97.3	99.5
Recall %	Walking	98.4	93.7	98.4
	Falling	89.7	98.7	100
	Picking up an object	100	100	100
Precision %	Walking	93.9	98.3	100
	Falling	98.6	95.1	98.7
	Picking up an object	91.3	100	100

$S_{H_k}(f,t)$ can be computed as $S_{H_k}(f,t) = |X_{H_k}(f,t)|^2$. The MDS $B_{H_k}^{(1)}(t)$ can be determined using the spectrogram as follows

$$B_{H_k}^{(1)}(t) = \frac{\int_{-\infty}^{\infty} f S_{H_k}(f,t)df}{\int_{-\infty}^{\infty} S_{H_k}(f,t)df}. \tag{2}$$

Since we use a commercial Wi-Fi NIC card for data acquisition, which measures the channel for each received data packet, the collected data suffers from several sources of error, such as noise, carrier frequency offset and sampling frequency offset (Yousefi et al. 2017). To mitigate these errors, we use various signal processing algorithms for data calibration, filtering, and denoising. This improves the quality of the estimated MDS. Note that a poor estimation of the MDS leads to a low accuracy in activity recognition. For each MDS sample, we extract a feature vector that is provided to the classification algorithm. The latter must determine the type of activity based on the feature vector.

4 EXPERIMENTAL RESULTS

In this section, we assess the performance of the proposed activity recognition system. First, the data is divided into training and test data sets representing 70% and 30% of the total data, respectively. A training example consists of the features extracted from a given MDS sample together with a label, which indicates the type of activity associated with the considered MDS sample. During the training phase, the classifier is exposed to labeled training data and learns the pattern associated with each activity. At the end of the training phase, the internal parameters of the classifier are tuned such that the trained algorithm can distinguish various activities with high accuracy. Subsequently, the performance of the trained classifier is evaluated using the test data set.

In our study, we compare the performance of three classification algorithms: CSVM, LDA, and KNN. The obtained results are illustrated in Table 1. This table shows that the CSVM algorithm achieves the best overall performance with an activity recognition accuracy of 99.5%, compared to 95.1% for KNN, and 97.3% for LDA. Table 1 provides the classification recall and precision of each algorithm for each activity. Note that the precision and the recall of the classification have two different meanings. The classification recall focuses on the actual activity and quantifies the percentage of successful classifications out of the *actual* samples belonging to a particular class. In contrast, the classification precision focuses on the predicted activity and indicates the percentage of correct classifications out of the samples *predicted* to belong to a certain activity. For instance, the classification recall for the activity falling for the CSVM algorithm is 100%. This implies that all the actual falls were correctly classified by CSVM. Whereas the classification precision of falling is 98.7% for CSVM, which means that 1.3% of the predicted falls are actually non-falling events.

The classification recall of the KNN classifier for the activities walking, falling, and picking up an object is equal to 98.4%, 89.7%, and 100%, respectively. Whereas for the LDA algorithm, the classification recall is 93.7%, 98.7%, and 100% for the activities walking, falling, and picking up an object, respectively. The CSVM algorithm reaches 100% recall for the activities falling and picking up objects and 98.4% recall for walking. The classification precision of the KNN classifier for the activities walking, falling, and picking up an object is equal to 93.9%, 98.6%, and 91.3%, respectively. While for the LDA algorithm, the classification precision is 98.3%, 95.1%, and 100% for the activities walking, falling, and picking up an object, respectively. The CSVM algorithm reaches 100% precision for the activities walking and picking up objects and 98.7% precision for falling.

5 CONCLUSION

In this paper, we have developed a HAR system, which consists of two building components: RF sensing and machine learning. The RF sensing component is a testbed used to collect RF data of indoor channels, while a human participant carries out three activities (walking, falling, and picking up an object). This RF testbed comprises two laptops acting as a transmitter and receiver. The transmitter laptop sends 1000 packets per second. At the receiver laptop, a software tool captures the CSI data in each received packet. This CSI data is a discrete version of the CTF sampled at 30 frequencies. The CSI data is collected from seven participants. Using signal processing algorithms, we reduce the noise in the CSI data and estimate the MDS associated with each CSI sample. A feature extraction algorithm is applied to each MDS sample to extract a feature vector. The data is then split into training and test data. The training data contains both the feature vector and a label indicating the type of the performed activity. We adopt a supervised learning approach, where the classifier is trained using the training data and then its performance is assessed using the test data. We evaluate the performance of three classification algorithms: CSVM, LDA, and KNN. Our experimental results show that the CSVM algorithm achieves the best performance with an overall accuracy of 99.5%, while LDA and KNN have an overall accuracy of 97.3% and 95.1%, respectively.

ACKNOWLEDGEMENT

This work is carried out within the scope of WiCare project funded by the Research Council of Norway under the grant number 261895/F20.

REFERENCES

Abdelnasser, H., M. Youssef, and K. Harras. 2015. "WiGest: A ubiquitous WiFi-based gesture recognition system." In *IEEE Conference on Computer Communications (INFOCOM 2015)*, Hong Kong, China, Apr., 1472–1480.

Boashash, B. 2015. *Time-Frequency Signal Analysis and Processing – A Comprehensive Reference*. 2nd ed. Elsevier, Academic Press, Amsterdam.

Chelli, A., and M. Pätzold. 2019. "A machine learning approach for fall detection and daily living activity recognition." *IEEE Access* 7: 38670–38687.

Halperin, D., W. Hu, A. Sheth, and D. Wetherall. 2011. "Tool release: Gathering 802.11N traces with channel state information." *ACM SIGCOMM Computer Communication Review* 41 (1): 53–53.

Sehairi, K., F. Chouireb, and J. Meunier. 2017. "Comparative study of motion detection methods for video surveillance systems." *Journal of Electronic Imaging* 26 (2): 26–29.

Wang, F., W. Gong, and J. Liu. 2019. "On spatial diversity in WiFi-based human activity recognition: A deep learning-based approach." *IEEE Internet of Things Journal* 6 (2): 2035–2047.

Wang, Y., K. Wu, and L. M. Ni. 2017. "WiFall: Device-free fall detection by wireless networks." *IEEE Transactions on Mobile Computing* 16 (2): 581–594.

Yousefi, S., H. Narui, S. Dayal, S. Ermon, and S. Valaee. 2017. "A Survey on Behavior Recognition Using WiFi Channel State Information." *IEEE Communications Magazine* 55 (10): 98–104.

Zou, H., Y. Zhou, J. Yang, H. Jiang, L. Xie, and C. J. Spanos. 2018. "DeepSense: Device-Free Human Activity Recognition via Autoencoder Long-Term Recurrent Convolutional Network." In *2018 IEEE International Conference on Communications (ICC)*, Kansas City, MO, USA, May, 1–6.

Zou, H., Y. Zhou, J. Yang, and C. J. Spanos. 2018. "Towards occupant activity driven smart buildings via WiFi-enabled IoT devices and deep learning." *Energy and Buildings* 177: 12 – 22.

Innovative and Intelligent Technology-Based Services for Smart
Environments – Ben Slama et al (eds)
© 2021 Taylor & Francis Group, London, ISBN 978-1-032-02030-3

Adapting resource utilization according to network needs using MEC, machine learning and network slicing

T. Abar
Cosim Lab, Supcom Tunisia, Carthage University, Tunisia

A. Ben Letaifa
Mediatron Lab, Supcom Tunisia, Carthage University, Tunisia

M. Abderrahim
Supcom Tunisia, Carthage University, Tunisia

S. Elasmi
Cosim Lab, Supcom Tunisia, Carthage University, Tunisia

ABSTRACT: Authors The 5G networks are designed to support a wide range of services that support the variety of requirements, performance and power utilization of services. The rapid increase in mobile network traffic, the number of connected devices, and the variety of applications being implemented require the efficient and dynamic management of network resources. Recently, several technologies have been proposed to improve 5G networks. Indeed, Mobile Edge Computing (MEC) presents a new opportunity to host applications close to users with a reduction in latency and improved performance. In addition, the concept of network slicing facilitates and the construction of multiple logical networks from a single physical infrastructure has been proposed, which optimizes flexible network evolution. In this article, we will seek to optimize the use of network resources, which are scarce resources, to meet the needs of customers and improve the delivery of services on demand by ensuring the smooth operation of services. In this sense, we start by studying the necessary needs in network resources using machine learning technology, then we apply the principle of network slicing based on the requirements of each service to ensure the dynamic and efficient management of network resources. Results show that our method improves resource utilization and ensures flexible, efficient and dynamic sharing of network resources.

Keywords: Mobile Edge Computing, Software Defined Networking, Machine learning, Smart Node, Network slicing, Orchestration, 5G.

1 INTRODUCTION

Next generation networks will be introduced to the market in the next few years. 5G opens new perspectives and allows the mixing of extremely diverse applications. 5G technology aims to facilitate the delivery of services and improve the communication network to the general public in order to find a balanced compromise between latency, speed and cost. The remarkable growth in the number of users, connected devices, and the growing number of cloudbased applications aims to improve the management and administration of network resources to meet a variety of service requirements. Several technologies are proposed to support the fifth-generation goal of providing better management of all networks and ensuring a dynamic, efficient and responsive distribution of resources. Software Defined Networking (SDN) and Network Functions Virtualization (NFV) have been emerged as key technologies for building the 5G system. In fact, SDN is a new approach characterized by the complete separation between the data plan and the control plan (Symeon 2020) which allows the programmability and automation of the network. NFV is a complementary

DOI 10.1201/9781003181545-14

technology to SDN. NFV (Ashwood-Smith et al. 2017) optimizes the use of network resources by virtualizing the functions usually implemented in dedicated equipment, which increases the deployment of new services with lower investment and operating costs. In addition, Mobile Edge Computing (Computing & Initiative 2014) is a key factor in the 5G network, it serves as a low latency communication that has one of the key features of 5G network. MEC facilitates the deployment of new services and applications closer to end users, reducing network costs and end-to-end delay. Indeed, this concept is currently attracting the attention of industry and society and it opens the door for researchers to improve the performance of the services offered. On the other hand, we cannot forget the technology of network slicing. Network slicing is a key factor for 5G network enhancement, it uses the concept of network virtualization that provides flexible and dynamic network management. This concept allows to share various heterogeneous virtual networks on a single physical network.

The concept of network slicing sliced admits several advantages over traditional networks, in fact, networks slices can dynamically evaluate based on the number of users and service requirements. In addition, network slicing enables service providers to deploy their applications in a flexible and timely manner to meet diverse service needs.

In this article, we took advantage of these technologies to propose a new architecture to improve the management of network resources. The objective of our work is to study and analyze the necessary network needs using machine learning algorithms and to apply the principle of network slicing in order to adapt the resource sharing policy dynamically according to network needs. The rest of the article is organized as follows. In section II, we define a context on the important technologies used. In section III, we describe our proposed architecture detailing the addition of the principle of network slicing in our architecture. Then we study our proposal on real data to show the importance of slicing to improve the management of network resources in section IV. And finally, we conclude and present our future work in section V.

2 BACKGROUND AND TECHNOLOGIES

In this section, we present a background on the different technologies used in this work. Indeed, MEC and network slicing present the main technology trends of the 5G network, they make it possible to meet the future needs of services with a good quality of services. In addition, machine learning technology is a means of analyzing and predicting network requirements based on historical data, which facilitates network management.

2.1 *Mobile Edge Computing*

Mobile Edge Computing (MEC) is a new standard technology introduced by the European Telecommunications Standard Institute (ETSI) in 2014 (Computing & Initiative 2014). MEC presents an important opportunity in the world of computer and telecommunication networks. MEC offers the ability to execute and process functionality within the Radio Access Network (RAN) and enables the deployment of real-time applications and services near mobile users with significantly reduced latency (Anta et al. 2017). Therefore, MEC will foster innovation by providing real-time and relevant information on the RAN as well as the core network to develop new applications. In the past, network traffic was transmitted to the core network to provide services to clients. However, the MEC concept aims to reduce the volume of a central traffic, thereby minimizing network overhead and freeing up latency.

In fact, the use of MEC servers simplifies the management of network traffic, reduces the congestion of the central network and facilitates real-time access to radio information. In (Sabella et al. 2016), the authors present an overview of the MEC framework and architecture, describe some examples of MEC deployment, and discuss the benefits and challenges of MEC for 5G. The MEC system offers several advantages for network operators, equipment and IT platform providers. The system aims to host new categories of end-user services and applications to reduce network delays and congestion. MEC technology is a key factor in the evolution of the fifth generation. It helps mobile network operators to expand their service portfolios and facilitates the flexibility and rapid delivery of applications near customers. In our work, this technology allows us to facilitate the collection of radio information in real time, which facilitates the identification of faster customer needs and the delivery of new services in a more agile way.

2.2 *Machine learning*

Machine learning is a type of artificial intelligence (AI) that allows machines and systems to learn and process tasks automatically without necessarily being programmed. Machine learning technology relies on the analysis, design and development of programs that enable a machine to handle difficult tasks by replacing them with more conventional and feasible algorithms. The machine learning technology is presented today in different domains, it can use for the type recognition of a document, the detection of fraudulent activities on the internet or the proposal of advertisements and results tailored to the expectations of customers on the search engine. It may also be in the medical sector where machines help diagnose cancer and in the banking sector to estimate a person's ability to repay a loan. The main objective of this technology is to reduce human intervention and enable machines to rely on historical data to predict meaningful outcomes of future developments and facilitate decision-making. The machine learning algorithms are classified into two categories:

— Supervised Learning: The supervised learning algorithms apply in two phases: the learning phase that requires historical data sets to train and, after the automatic learning phase, information processing is performed to predict future events.
— Unsupervised Learning: In contrast, unsupervised learning algorithms operate without a learning phase. They study how the system can solve problems from untagged data. There are several machine learning algorithms that define a prediction model and make the decision such as Decision trees, Neural networks, Gaussian Processes, Hidden Markov models, Dynamic Bayesian network, RandomForest, etc. In our work, we applied machine learning algorithms to facilitate knowledge of network traffic and thus ensure efficient management of network resources to meet customer requirements. For this, we used the weka framework which offers an interface for time series analysis. In fact, this interface makes it possible to generate forecasts for future events based on known past events. The purpose of our study is to take an idea of the traffic circulating in the network, based on information already collected and stored in a dataset, to exploit the network resources dynamically and efficiently.

2.3 *Network slicing*

Network slicing is a new concept introduced by Next Generation Mobile Network (NGMN) in (Flinck et al. 2017) to improve 5G networks. Network slicing has been identified today as the backbone that ensures the rapid evolution of 5G technology. The principle of network slicing is to share a single infrastructure to create multiple virtual ne works where each slice specified for a service or application. Figure 1 illustrates the concept of network slicing. This technology based on the concept of virtualization that allows to create virtual networks using software processes from a physical entity. The network slicing in 5G networks can include cutting the core network, radio access network (RAN) by logical abstraction of physical radio resources or end-user devices. RAN tranches and core network units must be dedicated to service requirements. Several researches have used the principle of network slicing emerging with SDN and NFV technologies. Indeed, in (Nikaein et al. 2015) the authors proposed a new 5G architecture based on SDN, NFV and cloud computing. They used the concept of a network store that can implement 5G dynamic network slicing. The authors in (Simon et al. 2016) have proposed a new 5G network architecture. This architecture aims to support the coexistence of heterogeneous services and facilitate the rapid deployment of new services in a flexible way. The use of SDN and NFV in the proposed architecture allows the sharing and orchestration of resources automatically between different services. The idea of the article

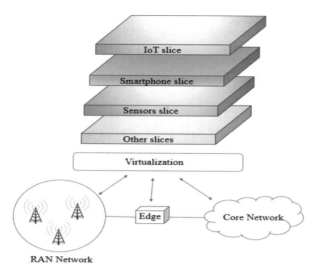

Figure 1. Network slicing.

(Rost et al. 2016) is to introduce an end-to-end multi-service 5G mobile network architecture using the principle of network slicing. In (Ksentini et al. 2017) the authors proposed a new architecture that can apply the network cutting with the abstraction of radio resources. The proposed architecture allows for flexibility and dynamism, using the FlexRAN concept, to impose network splitting in the RAN and adapt the resource allocation policy according to the needs of the slice. Network slicing allows the establishment of several autonomous logical networks on a common physical infrastructure.

This concept allows flexible and efficient management and administration of network resources. In Katsalis et al. 2017, a new architecture based on the concept of network partitioning is proposed to facilitate the delivery of custom network services and service models tailored to particular use cases. In our work, we proposed to apply the principle of network slicing to facilitate the sharing of resources to meet the desires of variety of applications running on the network. In fact, we have based on the result of the analysis and prediction performed using machine learning algorithms to ensure the dynamic and flexible distribution of network resources according to service needs. In the next section we will detail our proposed architecture model by specifying the role of Smart Node where we will add in this article the slicing principle to improve the resource management.

3   NETWORK MODEL AND PROPOSED FUNCTIONALITY

Knowing the rapid growth of mobile network traffic in recent years and the variety of requirements of services and applications present in the next 5G systems. To improve the management and administration of the mobile network with increasing network complexity, we proposed a new architecture in our previous work (Abderrahim et al. 2018). We can summarize here the principles of our proposal. We have benefited in our work from several technologies such as SDN, MEC and ML algorithms that have presented key factors in the evolution of the 5G network. Figure 2 presents our proposed architecture; the added value of our proposal is the integration of an intelligent entity that optimizes network operation.

This smart Node facilitates network management where it provides the following functions: i) radio information collection, ii) prediction of required resources, and iii) efficient resource allocation.

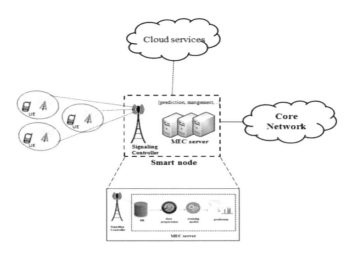

Figure 2.   Proposed architecture.

As shown in Figure 2, a signaling controller, which has a general view of the entire network, collects radio information from the base stations. This information is stored in a large dataset and updated periodically. Here we have benefited from MEC's principle that facilitates real-time access to radio information. Secondly, we take advantage of customer profiles and collected data (network parameters, client parameters, connection parameters, service parameters, connection times, etc.) to forecast the necessary network resource requirements. The idea of using machine-learning algorithms for prediction of needed needs facilitates network management and ensures a dynamic and intelligent allocation of network resources. To meet the different requirements of users and adapt the heterogeneity of 5G applications we propose to add in this article the principle of network slicing in the functionality of Smart Node. Indeed, network slicing is certainly one of the main factors allowing the evolution of 5G networks where our goal is to cut network resources in such ways where we can control networks in a dynamic and more agile way. Figure 3 shows the Smart Node in its new state. Firstly, the radio information control of all zones is done instantly by the signaling controller. The information collected contains all the necessary data to facilitate the prediction of the necessary needs.

The purpose of collecting radio data from all areas is to facilitate decision-making and prediction of the needs of each zone to ensure efficient and dynamic sharing of resources. Then, we applied the machine learning algorithms based on the collected information that gives a view of the network operation history. This phase consists in identifying the volume of traffic of each zone to ensure the necessary needs and guarantee the continuity and the good functioning of the applications. In fact, prediction simplifies resource sharing, flow control, and congestion control more efficiently. After the study and analysis of the old information collected, we have a forecast on the requirement of each zone. A slice orchestrator is responsible for orchestrating network resources by providing the necessary slices for each zone according to the prediction result. This method allows the network operator to deal with the problem of wasting network resources while respecting the variety of services and applications running on the network. In the following, we will try to show an example of our proposal by carrying out our study on a database that we had from a network operator in South Paris.

4   DISCUSSION

In this section, we discuss the results obtained when we apply our proposed approach. As shown in Figure 4, the network is divided into small areas. The Smart Node controls the radio information,

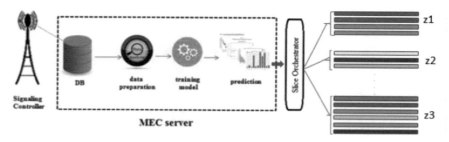

Figure 3. Smart node.

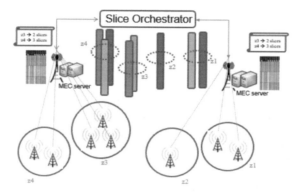

Figure 4. Functionalities of smart node.

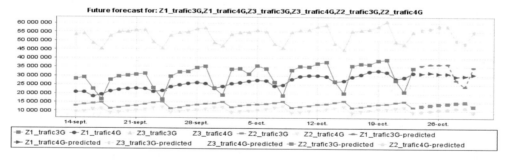

Figure 5. Traffic prediction.

it stocks it in a large dataset on which we will rely for the prediction of traffic volume of each part. Then, an investigation indicates the prediction result, performed at the MEC server level, is sent to the slice orchestrator who is responsible for assigning the slices needed for each zone. The numbers and nature of the slices vary according to the needs of each zone. In our case, we specified by the study of the traffics of a zone in the south of Paris. We applied the machine learning algorithm to predict the total volume of traffic per day based on the history of data stored in the dataset. We tried several learning algorithms where the application of the 'RandomForest' algorithm gives the most optimal results. Figure 5 shows the prediction result of total traffic volume per day in the near future. The algorithm trains on real data traffic obtained in the dataset to conclude the result of total volume of traffic in the following week. As shown in Figure 5, the total amount of traffic increases gradually over the course of the week and a sudden drop occurs during the weekend. It

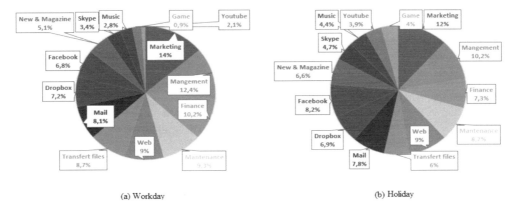

(a) Workday   (b) Holiday

Figure 6. Percentage of traffic for each service.

is obvious that in an industrial zone the network traffic decreases during the holiday and on the weekend because of the decrease of the services.

The purpose of our work is to take a general view of the network traffic and the characteristics of the services applied to help the operators to exploit their network resources. In the following we will try to dig deeper into the study and traffic analysis to properly identify the network slices needed for each area. The network resource requirements (bandwidth, frequency, latency time, QoS, etc.) vary according to several criteria such as the type of application, the nature of the area (forest, mountains, residential building, etc.), the power of use, mobility, etc. In our case, we were interested in analyzing the most frequent services in order to determine the characteristic and the requirements of each service.

Figure 6(a) shows the percentage of traffic for each service during a working day while Figure 6(b) shows the percentage of traffic during holiday. As shown, the traffic volume is not equal for all services. In addition, the use of the same service differs between the working day and the holiday. Indeed, during the weekend, some traffic is reduced in one area while other services tighten more powerful in another area. Taking for example gaming applications that do not exceed 1% of the total traffic volume during the week but can reach up to 4% in the weekend. The dynamic network resource sharing offered by the slice orchestrator saves a lot in terms of response time and resource utilization. Indeed, if we fix the slices offered for the game services (for example we set the slices for use during the workday), it causes problems of service interruption or delay to respond to customer requests. So, the dynamic orchestration of the slices makes it possible to find a compromise to use the network resources dynamically while guaranteeing the quality of service. In general, the requirements of services vary from one service to another, in fact there are services tolerated in time, others tolerated in error, others require high capacity or are characterized by high mobility, etc. Service operators always seek to improve the delivery of their services by respecting the requirement of each service. The concept of network slicing allows to meet various performance requirements like low latency, broadband, high mobility, etc. Indeed, network slicing provides each service with a specific logical network on a single physical network infrastructure. Thus, with the new network slicing technology, we can provide better performance for logical networks than for large networks. The idea behind our work helps network operators benefit from the exploitation and use of network resources that today represents a problem in the face of rapid increases in network traffic and limited network resources. Indeed, the prediction phase allows us to provide a general view of the network traffic to prepare the necessary needs to meet the requirements of users ensuring continuity of service. Then, the use of network slicing which makes it possible to optimize network management and evolution enables the deployment of a broad portfolio of services on a single physical network.

5 CONCLUSION

Fifth generation mobile communications have been developed to meet multiple types of services. In the last years, mobile data traffic is the largest part of network traffic. This traffic differs from one zone to another and varies from workdays to the weekend. For this, dynamic management of resources becomes a very important requirement for the 5G network that characterizes the variety and heterogeneity of services. In this article, we have improved our previously proposed architecture to optimize the sharing of resources to meet the different needs of customers. First, we began by prediction and analysis of applications running on the network, applying machine learning algorithms, to prepare the appropriate ground. Next, we proposed to use the network slicing principle to adapt network slices according to network needs. This method improves resource utilization and ensures flexible, efficient and dynamic sharing of network resources.

The proposed method in this paper aims to improve the network performances that enhance the user satisfaction and the quality of experience QoE by facilitating the delivery of services and finding a balanced compromise between latency, speed and cost. We aim to introduce this approach in our future work in order to enhance the QoE of video streaming service. The idea will be based on the combination of the network slicing described in this paper and the use of network nodes to cache user-requested content. This proposal can easily decrease the end to end delay and the user will be satisfied with the service.

REFERENCES

Abderrahim, M., Letaifa, A. B., Haji, A., & Tabbane, S. (2018, May). How to use MEC and ML to improve resources allocation in SDN networks?. In *2018 32nd International Conference on Advanced Information Networking and Applications Workshops (WAINA)* (pp. 442–447). IEEE.

Ashwood-Smith, P., Mohammadi, M. A. A., & Evelyne, R. O. C. H. (2017). *U.S. Patent No. 9,847,915*. Washington, DC: U.S. Patent and Trademark Office.

Computing, E. M. E., & Initiative, I. (2014). Mobile-edge computing: introductory technical white pa per. *ETSI: Sophia Antipolis, France, 1*–36.

Huang, A., Nikaein, N., Stenbock, T., Ksentini, A., & Bonnet, C. (2017, May). Low latency MEC framework for SDN-based LTE/LTE-A networks. In *2017 IEEE International Conference on Communications (ICC)* (pp. 1–6). IEEE.

Katsalis, K., Nikaein, N., Schiller, E., Ksentini, A., & Braun, T. (2017). Network slices toward 5G communications: Slicing the LTE network. *IEEE Communications Magazine, 55*(8), 146–154.

Ksentini, A., & Nikaein, N. (2017). Toward enforcing network slicing on RAN: Flexibility and resources abstraction. *IEEE Communications Magazine, 55*(6), 102–108.

Nikaein, N., Schiller, E., Favraud, R., Katsalis, K., Stavropoulos, D., Alyafawi, I., ... & Korakis, T. (2015, September). Network store: Exploring slicing in future 5G networks. In *Proceedings of the 10th International Workshop on Mobility in the Evolving Internet Architecture* (pp. 8–13).

Papavassiliou, S. (2020). Software Defined Networking (SDN) and Network Function Virtualization (NFV).

Rost, P., Banchs, A., Berberana, I., Breitbach, M., Doll, M., Droste, H., ... & Sayadi, B. (2016). Mobile network architecture evolution toward 5G. *IEEE Communications Magazine, 54*(5), 84–91.

Sabella, D., Vaillant, A., Kuure, P., Rauschenbach, U., & Giust, F. (2016). Mobile-edge computing architecture: The role of MEC in the Internet of Things. *IEEE Consumer Electronics Magazine, 5*(4), 84–91.

Simon, C., Maliosz, M., Bíró, J., Gerö, B., & Kern, A. (2016, June). 5G exchange for inter-domain resource sharing. In *2016 IEEE International Symposium on Local and Metropolitan Area Networks (LANMAN)* (pp. 1–6). IEEE.

Sprecher, N. (2017). Internet Engineering Task Force H. Flinck Internet-Draft C. Sartori Intended status: Informational A. Andrianov Expires: January 4, 2018 C. Mannweiler. Network.

Innovative and Intelligent Technology-Based Services for Smart Environments – Ben Slama et al (eds)

Toward a modular architecture for fault-tolerance in complex systems

F. Lassoued & R. Bouallegue
Innov'COM Laboratory, Higher School of Communications of Tunis, Sup'Com, Carthage University, Tunisia

ABSTRACT: Throughout this paper, we implement a distributed application based on component model abstraction. We focus on the naval battle game as an example to implement modular architecture using a Fractal component model. We show that a modular architecture allows an automatic failure detection and automatic reparation. This work is classified in the context of automatic computing areas which attracts today researchers and industrialists.

1 INTRODUCTION

Distributed computing is generally acknowledged as a common term that refers to an infrastructure that involves the integration and collaboration among heterogeneous and geographically distributed resources. It has to do with devising algorithms for a set of processes that seek to achieve some form of cooperation. Besides executing concurrently, some of the processes of a distributed system might stop operating, for instance by crashing or by being disconnected, while others might stay alive and keep operating. This very notion of partial failure is a characteristic of a distributed system.

In fact, many programs that we use today are distributed programs. Simple daily routines, such as reading e-mail or browsing the web, involve some form of distributed computing. However, when using these applications, we are typically faced with the simplest form of distributed computing: client-server computing. In client-server computing, a centralized process, the server, provides the service for many remote clients. The clients and the server communicate by exchanging messages, usually following request-reply interaction form. For instance, in order to display a web page to the user, a browser sends a request to the web server and expects to obtain a response with the information to be displayed.

The core difficulty of distributed computing, namely, achieving a consistent form of cooperation in the presence of partial failures, may pop up even by using this simple form of interaction. Going back to the browsing example, it is reasonable to expect that the user continues surfing the web (by automatically being switched to other sites) if the site it is consulting fails, it is even more reasonable that the server process keeps on providing information to the other client processes, even when some of them fail or get disconnected.

Actually, a real distributed system would have parts following client-server interaction patterns and other parts following a multiparty interaction pattern. For instance, when a client contacts a server to obtain a service, it may not be aware that, in order to provide that service, the server itself may need to request the assistance of several others servers, with whom it needs to coordinate to satisfy the client's request.

Therefore, high availability and fault tolerance has become a feature for such systems in order to make the communication between these resources efficient and coherent and keep the system or the application in a consistent state.

Software fault tolerance is achieved through the technique called replication. Due to replication, several software replicas are executed at the same time. If one or several of them fail, other still provides the service. Nevertheless, the faults are not totally struck out and the challenge is usually

DOI 10.1201/9781003181545-15

for the processes that are still alive and keep operating, or connected to the majority of processes, to synchronize their activities in a consistent way. Moreover, it's crucial to detect the failure occurrence and repair the crashed component.

Indeed, software replication is often implemented using a group communication system, which provides communication primitives with various semantics and greatly simplifies the development of highly available and fault tolerant services. One such primitive is called a total order broadcast. It ensures that messages sent to a set of processes are delivered by all those processes in the same order.

On the other hand, to repair the failure we need to combine the work of two reviews; algorithmic review and failure correction review. The first one aims to detect the fault occurrence in order to keep the system in a consistent state when the second presents the need for a flexible architecture that make the correction task possible.

The problems listed above motivate the need for fault tolerance technique to well-build a distributed system. Therefore, during this paper, we focus on fault-tolerant systems, in particular on two concepts of fault tolerance: algorithmic and architectural. In regard to architectural review, we propose to implement a modular architecture for a game, the naval battle game, as a fault-tolerance illustration on a distributed system using the fractal component model.

This work consists on the design and development of a fault-tolerant distributed system based on the fractal component model. Indeed, we introduce the notion of the component abstraction model, in particular, the fractal component model. Finally, we implement the naval battle game as a fault tolerance illustration on a distributed system using the Fractal component model.

2 COMPONENT MODEL ABSTRACTION

2.1 Definition of a component

"A component is a unit of composition with contractually specified interfaces and context dependencies only. A software component can be deployed independently and is subject to composition by third parties" (Szyperski 1997).

By enforcing a strict separation between interface and implementation and by making software architecture explicit, component-based programming can facilitate the implementation and maintenance of complex software systems. Coupled with the use of meta-programming techniques, component-based programming can hide to application programmers some of the complexities inherent in the handling of non-functional aspects in a software system, such as distribution and fault-tolerance, as exemplified by the container concept in Enterprise Java Beans (EJB), CORBA Component Model (CCM) or Microsoft .Net.

2.2 Fractal component

The Fractal component model introduces a notion of components endowed with an open set of control capabilities. In other terms, components in fractal are reflective, and their reflective capabilities are not fixed in the model but can be extended and adapted to fit the programmer's constraints and objectives.

The fractal component model (Figure 1) is a general component model which is intended to implement, deploy and manage complex software systems, including in particular operating systems and middleware. This motivates the main features of the model:

- Composite component: to have a uniform view of application at various levels of abstraction.
- Shared component: to model resources and their sharing while maintaining component encapsulation.
- Introspection capabilities: to monitor a running system.
- Re-configuration capabilities: to deploy and dynamically configure of component a system.

90

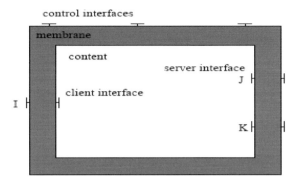

Figure 1. The fractal component.

In order to allow programmers to tune the reflective features of components to the requirements of their applications, fractal is defined as an extensible system. Control features of components are not predetermined in the model, rather the model allows for a continuum of reflective features or levels of control, ranging from no-control to full-fledged introspection and intercession capabilities.

2.2.1 *Components and binding*

Fractal component is a run-time entity that is encapsulated, and that has a distinct identity and that supports one or more interfaces. An interface is an access point to a component that implements an interface type (i.e. a type specifying the operations supported by the access point). Interfaces can be of two kinds: sever interfaces, which correspond to access points accepting incoming operation invocations, and client interfaces, which correspond to access points supporting operation invocations.

A composite binding is a communication path between an arbitrary numbers of components (stubs, skeletons, adapters…). A binding is a normal Fractal component whose role is to mediate communication between other components. The binding concept corresponds to the connector concept that is defined in other component models. Note that, except for primitive bindings, there is no predefined set of bindings in fractal.

In fact bindings can be built explicitly by composition, just as other components. The fractal model thus provides two mechanisms to define the architecture of an application: bindings between component interfaces, and encapsulation of a group of components in a composite.

Communication between fractal components is only possible if their interfaces are bound. Fractal supports both primitive bindings and composite binding. A primitive binding is a binding between one client interface and one server interface in the same address space, which means that operation invocations emitted by the client interface should be accepted by the specified server interface. A primitive binding is called that way because it can be readily implemented by pointers or direct language references (e.g. Java references).

The fractal model is endowed with an optional type system (some components such as base components which need not adhere to the type system). Interface types are in pairs (signature, role). Component types reflect the different interface types that a component can bear.

A fractal component (see figure above) can be understood generally as being composed of a membrane, which supports interfaces to introspect and reconfigure its internal features and content. The membrane of a component can have external and internal interfaces. External interfaces are accessible from outside the component, while internal interfaces are only accessible from the component's sub-components.

The membrane of a component is typically composed of several controllers. Typically, a membrane can provide an explicit and causally connected representation of the component's sub-components and superpose a control behavior to theirs. Controllers can also play the role of interceptors.

Interceptor objects are used to export the external interfaces of a sub-component as an external interface of the parent component. They intercept the oncoming and out coming operational invocations of an exported interface and they can add additional behavior to the handling of such invocations (e.g. pre and post-handlers).

Each component membrane can thus be seen as implementing particular semantics of composition for the component's sub-components. Controller and interceptors can be understood as meta-objects or meta-groups (as they appear in reflective languages and systems).

2.2.2 *Control levels*

At the lowest level of control, a fractal component is a black box that does not provide any introspection or intercession capability. Such components, called base components are comparable to plain objects in an object-oriented programming language such as Java. Their explicit inclusion in the model facilitates the integration of legacy software.

At upper levels of control, a fractal component exposes (part of) its internal structure. A fractal component is composed of a membrane, which provides external interfaces to introspect and reconfigure its internal features (called control interfaces), and a content, which consists in a finite set of other components (called sub-components).

The fractal model allows for arbitrary (including user defined) classes of controller and interceptor objects. This is the main reason behind the denomination 'open component model'. The fractal specification, however, contains several examples of useful forms of controllers, which can be combined and extended to yield components with different reflective features. The following are examples of controllers.

- Attribute controller: An attribute is a configurable property of a component. A component can provide an AttributeController interface to expose getter and setter operations for its attributes.
- Binding controller: A component can provide the BindingController interface to allow binding and unbinding its client interfaces to server interfaces by means of primitive bindings.
- Content controller: A component can provide the ContentController interface to list, add and remove sub-component in its contents.
- Life-cycle controller: A component can provide the LifeCycleController interface to allow explicit control over its main behavior phases, in support for dynamic reconfiguration. Basic lifecycle methods supported by a LifeCycleController interface include methods to start and stop the execution of the component.

2.2.3 *Implementation*

The JULIA framework supports the construction of software systems with FRACTAL components written in Java. The main design goal for JULIA has been to implement a framework to program fractal component membranes. In particular, it provides an extensible set of control objects, from which the user can freely choose and assemble the controller and the interceptor objects in his or her own manner, in order to build the membrane of a fractal component.

The second design goal has been to provide a continuum from static configuration to dynamic reconfiguration. The last design goal was to implement a framework that can be used on any JVM and/or JDK.

A fractal component is generally represented by many Java objects, which can be separated into three groups (see Figure 2):

- The objects that implement the component interfaces are in white in (Figure 2) (one object per component interface; each object has a reference to an object that really implements the Java interface, and to which all method calls are delegated; this reference is null for client interfaces; for server interfaces it can refer to an interceptor or an object of the content part).
- The objects that implement the membrane of the component, in gray and light in (Figure 2) (a controller object can implement zero or more control interfaces).
- The objects that implement the content part of the component (not shown in Figure 2).

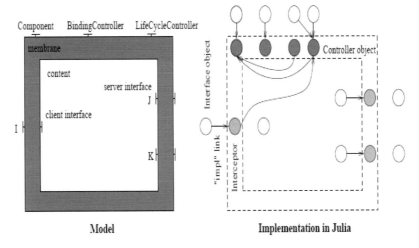

Component   BindingController   LifeCycleController

membrane

content

server interface

J

client interface

I

K

Interface object

"impl" link

Interceptor

Controller object

Model   **Implementation in Julia**

Figure 2.   Implementation with the framework JULIA.

Moreover, the objects that represent the membrane of a component can be separated into two groups: the objects that implement the control interfaces (in gray in the fig), and (optional) interceptor objects (in light gray) that intercept incoming and/or outgoing method calls on non-control interfaces. These objects implement respectively the Controller and the Interceptor interfaces. Each controller and interceptor object can contain references to other controller/interceptor objects.

2.2.4 *Deployment:*

We present two aspects of deployment: Architecture Description Language (ADL) and the ClassLoader mechanism:

- Fractal ADL

Fractal provides an architecture description language composed of two parts: a language based on XML and a factory that allows dealing with the definitions giving by this language. The Fractal ADL language is extensible: it composed of a set of modules describing all application aspects. For instance, Figure 4 below presents the ADL definition of the component in Figure 3:

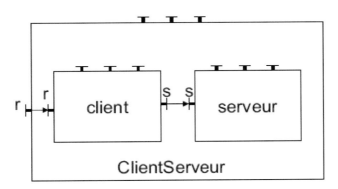

r   r

r

client

s   s

serveur

ClientServeur

Figure 3.   ClientServer example.

```
<definition name="ClientServeur">
<interface name="r" role="server" signature="java.lang.Runnable"/>
<component name="client">
    <interface name="r" role="server" signature="java.lang.Runnable"/>
    <interface name="s" role="client" signature="Service"/>
    <content class="ClientImpl"/>
</component>
<component name="serveur">
    <interface name="r" role="server" signature="java.lang.Runnable"/>
    <interface name="s" role="client" signature="Service"/>
    <content class="ClientImpl"/>
</component>
<binding client="this.r" server="client.r"/>
<binding client="client.s" server="serveur.s"/>
</definition>
```

Figure 4.   The ClientServer ADL description.

In fact, we distinguish a composite component which is ClientServeur and two sub-Components.

- The class loader

We can load a class using MySimpleClassLoader like this:

```
ClassLoader cl = new MySimpleClassLoader();
Map context = new HashMap();
context.put("classloader", cl);
Factory f = FactoryFactoty.getFactory();
Object c = f.newComponent("ClientServeur", context);
```

Figure 5.   The ClientServer ClassLoader.

2.3 *Summary*

Fractal allows building modular and explicit architectures. The Fractal systems can be dynamically reconfigured for failure reparation for instance as we show it in this work.

3   APPLICATION: IMPLEMENTATION OF THE BATTLE GAME

3.1 *Description*

The classical naval battle game involves at least two players and one coordinator. In the beginning, each player has to put a number, n, of ships onto this board. The number of ships is fixed. After that the game begun and each player tries to guess the coordinates of the other player. So that until one of them guess all the ship coordinates and being the winner. In regard to the coordinator role, it keeps a copy of the two boards, in order to make the task of cheat impossible. To control the game, the coordinator serves as an intermediary in the request-reply interaction (Boolean saying if the attempt is true or false). The functional view is described in Figure 6.

Indeed, we extend the classic game of the naval battle to tolerate the failure of the coordinator so that the information is providing even when some of them fail. As a result, the fault-tolerance motivates the need for synchronization between two coordinators to update their boards.

In fact, the administrator component sends a ping to each coordinator and if the timeout expired without receiving a reply from the coordinator concerned, it will be suspected by the administrator.

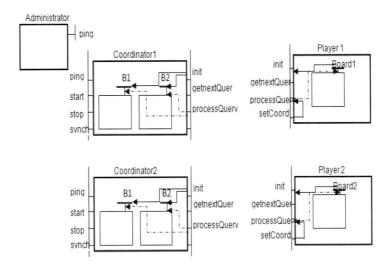

Figure 6.   Functional view of the battle game.

In this case, the administrator sends a request to inform the players about the new coordinator. Several cases are interesting to highlight (see Figure 6):

- The administrator is implemented by the component "administrator".
- The players are implemented by the component "player 1" and "player 2." Each component "playeri" has the listed interfaces:
  - init (nb; b): where nb is the number of the ships to be placed in each board while b give an information about the given board.
  - getnextQuery (null; coord): it returns a proposal.
  - processQuery (coord;null): it allows to check the proposal "coord" and return true or false.
  - setCoord (coord): used by the administrator to inform the player about the correct coordinator.

3.2   Description

3.2.1   Definitions

As the development of informatics systems becomes more complex, the need for a design method-ologies increases and several methodologies have been published during the last three decades such as Booch, Classe Relation, Fusion, HOOD, OMT, OOA, OOD, OOM, OOSE etc. Each method is defined by specific process and semantics. Hence, on 1997, the UML (Unified Modeling Lan-guage) allows the unification and the standardization of the concepts of three dominating object methods:

- OMT (Objet Modeling Technique).
- OOD (Object Oriented Design).
- OOSE (Object Oriented Software Engineering).

The UML based on a set of diagrams organized, into two views: the static view and the dynamic view. Going back to our application, it is interesting to use the UML.

3.2.2   Modeling

Going back to the application, it is reasonable to use UML to model the architecture. We start by the user cases in which we define the functionalities of the application. As we describe the scenario in 3.1, we introduce the user cases directly (see table 1):

Table 1. The user cases.

Providing a game	– Coordinator – Player	– Creation, board initialization, coordination, Synchronization, – Coordinate guess, verification
Failure detection	– Administrator – Coordinator	– Ping request, set coord – Ping reply
Failure repair	– Administrator	– Automatically replace the faulty – coordinator

- Providing the game

Table 2. The user case "Providing the game".

Identification content

Title	Providing the game
Purpose	Implement the naval battle game
Actor(s)	Coordinator and player

Table 3. The description of the user case "Providing the game".

Sequences description

Initial state: starting the coordinator1 at the beginning of each by the administrator scenario

Sequencess:
- Use the init interface by the coordinator1 to start the game and initialize the two boards (put n shops on the board)
- The coordinator keep a copy of each board
- Starting the attempts to guess the positions of the ships (getnextQuery and ProcessQuery interfaces)
- Update the boards states either for the running coordinator and the other one (Synchronize)
- Until the score of one of the two player reaches n so that he is the winnere and the game stopped

End State
 The winner

- Failure detection and repair

Table 4. The user case "Failure detection and repair".

Identification content

Title	Failure detection
Purpose	Provide the information even when the coordinator fails or being disconnected
Actor	Administrator and Coordinator

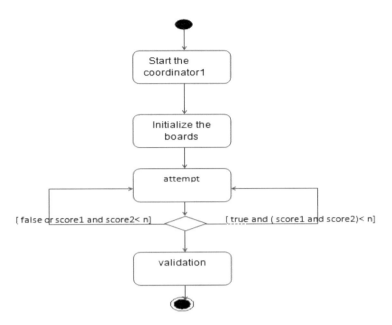

Figure 7. The activity diagram of the user case "providing the game".

Table 5. The description of the user case "Failure detection and repair".

Sequences description

Initial state: running the game

Sequences:
– The administrator sends a ping request at each ping frequency to the running coordinator.
  If the timeout expires and it don't receive a ping reply from the coordinator then the
  administrator uses the setCoord interface to inform the players about the new coordinator
  without need to stop the application
– Therefore, we implement a synchronization mechanism between the two coordinators to still
  provide the information even the coordinator fails

Final state: Failure detected

3.3 *Architecture*

The topic is the architecture of a distributed system, so that, it's reasonable to implement a distributed architecture of the game. In this case, description of the architecture is given in Figure 9 where each component is located on a different JVM in the sense that this architecture will be running in 5 JVM.

We implement the Fractal Component Model. As we describe above, we have four different components; the administrator component, the player component, the coordinator component which are the sub-components and the "BatailleNavale" component as a composite component. Each component has the appropriate interfaces. The architecture is given in Figure 10.

The administrator component starts the game by sending the request (using the interface Set-Coord(coord)) to the players and then executes the coordinator1 start method. The init interfaces allows initializing the board. Then the players can guess the coordinates of their opponents. Due to

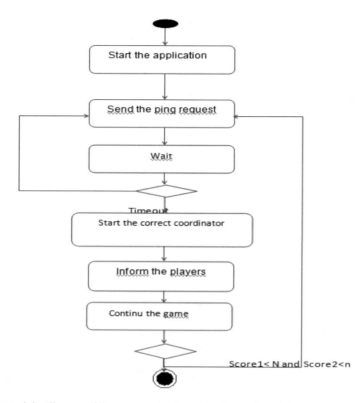

Figure 8.   The activity diagram of the user case "failure detection and repair".

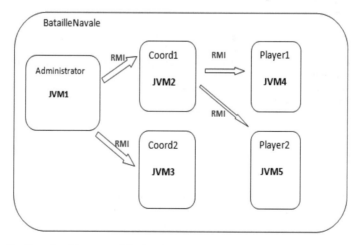

Figure 9.   The distributed architecture of the battle game.

this component abstraction, the architecture of the application is more flexible and the failure of a coordinator doesn't stop the execution of the application.

- The Fractal ADL:

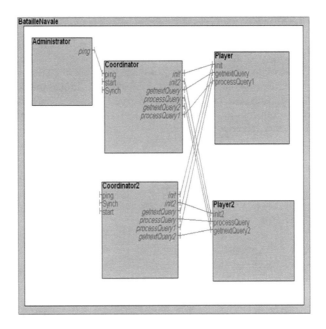

Figure 10. The fractal naval battle architecture.

In regards to the deployment, this is an extract from the ADL file describing the whole architecture.

```
<! DOCTYPE definition PUBLIC "-//objectweb.org//DTD Fractal ADL 2.0//EN"
"classpath://org/objectweb/fractal/adl/xml/standard.dtd">

<definition name="BatailleNavaleType"/>

<definition name="CoordinatorType">
  <interface name="ping" role="server" signature="Ping"/>
  <interface name="start" role="server" signature="StartItF"/>



  <interface name="init" role="client" signature="Init"/>
  <interface name="init2" role="client" signature="Init2"/>
  <interface name="getnextQuery" role="client" signature="GetNextQuery"/>
  <interface name="processQuery" role="client" signature="ProcessQuery"/>
  <interface name="getnextQuery2" role="client" signature="GetNextQuery2"/>
  <interface name="processQuery1" role="client" signature="ProcessQuery"/>
</definition>

<definition extends="CoordinatorType" name="Coordinator"/>

<definition name="PlayerType">
  <interface name="init" role="server" signature="Init"/>
  <interface name="getnextQuery" role="server" signature="GetNextQuery"/>
  <interface name="processQuery1" role="server" signature="ProcessQuery"/>
</definition>

<definition extends="PlayerType" name="Player">
  <content class="PlayerImpl"/>
</definition>
```

Figure 11. ADL file.

- The Administrator component:

The Administrator component has a client interface denoted by "ping" so that the main role of this component is to control the coordinator. As in the extract from the ADL file describing his behavior, it defines two control interfaces, in particular two attributes: Ping Frequency and Timeout. So, this component has to send a ping request every 1500s and wait for a reply. If the timeout equal to 1000 expires, the coordinator concerned will be suspected (see Figure 12 and Figure 13). The class AdminImpl defines the behavior of this component and his interaction with coordinator component and player component.

```
<definition name="AdministratorType">
    <interface name="ping" role="client" signature="Ping"/>
    </definition>

    <definition extends="AdministratorType" name="Administrator">
    <content class="AdminImpl"/>

    <attributes>
    <attribute name="PingFrequency" value="1500"/>
    <attribute name="Timeout" value="1000"/>
    </attributes>
</definition>
```

Figure 12.   An extract from Administrator ADL file.

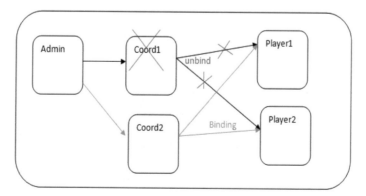

Figure 13.   The coordinator1 fails.

- The Coordinator component:

This component has both client interfaces and server interfaces. It implements the semantic of the game. The running coordinator has to, at each interaction with a player component,;eep a copy of the boards. It implements the LifeCycleController interface.

- The Player component:

The player component is defined as a board. The server interface "init" consist in the initialization of the board. On the other hand the interface getnextQuery allows the player to attempt a pair of coordinates while the server interface "processQuery" allows the player to check the attempt of his opponent.

4 DISCUSSION

The component abstraction model is the key aspect of the distributed systems. Indeed, as an example, the main goals of the fractal component model are to implement, deploy and manage (i.e. monitor and dynamically reconfigure) complex software systems.

The Fractal component model is not defined as a fixed specification that all fractal components must follow, but rather as an extensible system of relations between well-defined concepts and corresponding APIs that fractal components may or may not implement, depending on what they can or want to offer to other components.

At the lowest level of control, a fractal component is a runtime entity that does not provide any control capability to other components, and is therefore like an object (such a component can be used in only one way, namely by calling methods on it). In fact, an object is a Fractal component without any control capability.

While, at the next level of control capability, which can be called the external "introspection" level, a fractal component can provide a standard interface that allows one to discover all its external interfaces or, in other words, its boundaries (like an object, a Fractal component can provide several interfaces).

On the other hand, at the next level of control capability, which can be called the "configuration" level, a fractal component can provide control interfaces to introspect and modify its content, i.e. what is inside its boundary. In the fractal model, this content is made of other fractal components, called its subcomponents, bound together through bindings. A fractal component can therefore choose to provide or not an interface to control the set of its subcomponents, the set of bindings between these subcomponents, and so on.

As a result of this modular and extensible organization (anyone is free to define their own control interfaces, in order to provide new introspection and intercession capabilities), and given the fact that the fractal component model is not tied to a specific language, fractal components can be used in very different situations, from operating systems to middleware platforms, from graphical user interfaces to information systems, and from highly optimized but un-reconfigurable configurations, to less optimized but highly dynamic and reconfigurable systems or applications. This approach hopefully simplifies the task of the distributed programming, and resolve the major problems, such as administration and fault tolerance, that it difficult to deal with especially for a system.

5 CONCLUSION

We have presented the fractal component model and its Java implementation, JULIA. Fractal is open in the sense that fractal components are endowed with extensible set of reflective capacities (controller and interceptor objects). JULIA consists in small run-time library. Using this abstraction model, we have implemented a fault-tolerant application which is the naval battle game either for a local architecture in the sense that all components are in the same JVM and for a distributed architecture with 5 JVM. We have illustrated an automatic repair scenario where a component is detected as faulty and is automatically replaced with another.

REFERENCES

Chandra. T. D and Toueg. S. Unreliable failure detectors for reliable distributed systems. Journal of ACM, 43(2):225–267, 1996.

Ekwall. R. and Shipper. A. Modeling and Validating the Performance of Atomic Broadcast Algorithms in High Latency Networks.Europar 2007.

Ekwall. R., Schiper. A, and Urbàn. P. Token-based atomic broadcast using unreliable failure detectors. In Proc. of the 23rd Symposium on Reliable Distributed Systems (SRDS 2004), Florianopolis, Brazil, Oct. 2004.

Guerraoui. R, Levy. R. R, Pochon. B, and Quéma. V. High Throughput Total Order for Cluster Environments. In IEEE International Conference on Dependable Systems and Networks (DSN 2006), June 2006.

Mostefaoui. A and Raynal. M. Solving Consensus using Chandra-Toueg's Unreliable Failure Detectors: A Synthetic Approach. In 13th. Intl. Symposium on Distributed Computing (DISC'99). Springer Verlag, LNCS 1693, September 1999.

Innovative and Intelligent Technology-Based Services for Smart Environments – Ben Slama et al (eds)
© *2021 Taylor & Francis Group, London, ISBN 978-1-032-02030-3*

Design and realization of a broadband printed dipole array for the radio-localization of mobile phones from drones

J.M. Floc'h & I. Ben Trad
IETR, INSA Rennes, Rennes, France

A. Ferreol & P. Thaly
Thales Communications & Security, Paris, France

ABSTRACT: We present in this paper the study and characterization of a network of antennas conceived within the framework of the SOSPEDRO project (Surveillance de zones Sinistreìes et de Personnes par Drone). The objective of this project is to help rescue forces in their missions to monitor areas and people in distress using a drone-mounted system. We have developed a circular antenna array of 8 wide-band antennas with polarization diversity in order to cover the frequency range from 400 to 2400 MHz (PMR, GSM 800/900/1800/1900, UMTS and LTE). The search and radio-localization of people in a hostile environment will be carried out using the emission of their mobile phones.

1 INTRODUCTION

The proposed radio-localization system will be mounted on a mini drone which is easily deployable. Emergency professionals engaged in intervention zones will be able to monitor the areas of interest (coastal areas, flooded areas, borders, forest fires, traffic, delinquency, plane crash, terrorist attacks, etc.). The difficulties to solve in the design of this array are multiple: it must have a good linear polarization in all bandwidths, the array doesn't have a weight greater than 1 Kg, the array is circular with a 40 cm diameter, have a low back radiation. We use different criteria in order to define the positions of dipoles. We use a genetic algorithm to find the good positions. We realize three different arrays in order to compare the obtained results.

2 THE ELEMENTARY ANTENNA: DIPOLE

We have studied and developed a circular network of 8 identical dipoles and wide bands in order to cover (S11 <−6 dB) the bands PMR, GSM 800/900804/5000/ 1800/1900, UMTS and LTE (Stec et al. 2014; Liu 2016; Cherntanomwong et al. 2007; Pralon et al. 2015). Figure 2 shows the proposed antenna being able to radiate from 400 MHz–3GHz (for S11 <−6 dB). Now we must adapt the antennas at low frequencies PMR, LTE (700–900 MHz) and GSM 800/900. This is due to the size of the sensors. We add a coaxial attenuator R411.803.124 for this purpose, as it is shown in figure 4. This technique is well known in order to reduce the reflection coefficient at low frequencies (we use a 3 dB attenuator). The measured 3D radiation patterns of the elemental antenna shown in figure 3 show a uniform forward radiation (in the direction of the ground) and a weak back radiation (in the direction of the drone). Although gain is not a primary criterion, at 900 MHz the antenna has a maximum gain measured at Gmax = 1.19 dBi and at 2170 MHz Gmax = 4.56 dBi and can reach Gmax = 5.77 dBi at 2400 MHz.

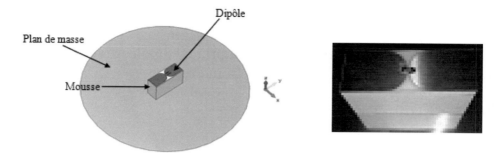

Figure 1.   Single antenna.

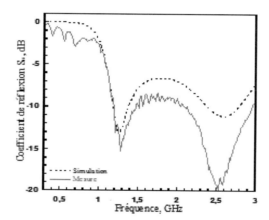

Figure 2.   Simulated and measured reflection coefficients.

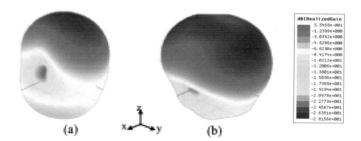

Figure 3.   Measured 3D radiation patterns at 900 MHz (a) and 2170 MHz (b).

In this project, we fabricate 3 different arrays in order to compare the performance of these arrays. It is very important because these arrays have very different structures. The dipoles are placed arbitrarily on the ground plane and we use a genetic algorithm to place the dipoles and we use for this purpose a certain number of criteria (you can find these criteria in the two communications (Ben Trad et al. 2017) and (Floc'h et al. 2018)).

The orientation and position of each antenna have been optimized to obtain good polarization diversity with the smallest possible backward radiation. The goal is to be able to capture the signals emitted by mobile phones in the areas of interest and thus accurately determine the position of the wanted people. The choice of the basic antenna was focused on the dipole shown in figure 1.

The dipole is printed on the substrate FR4 of dimensions $83 \times 30 \times 1.6$ mm with permittivity $\varepsilon_r = 4.4$. The foam serves as a support for maintaining the dipole at 47.7 mm from the 40 cm diameter ground plane.

We define 3 criteria for the array determination:

- Figure 4: Array Gain: an example at 450 MHz for 2 polarizations. We try to minimize the radiation in the blades direction (under -7 dB).
- Figure 5: Depolarization criterion in a Theta direction $\Psi(\phi, \Theta)$: It is the scalar product of the polarization vector of an antenna with a reference.
- Figure 6: the array similarity for two polarizations at 900 MHz.

We define the similarity as the difference between the theoretical array and the results obtained through the simulation or measurements. The difference between the two is due to the finite size of the ground plane, the edge effect and the coupling effect between dipoles.

We try to obtain the similarity close to 1.

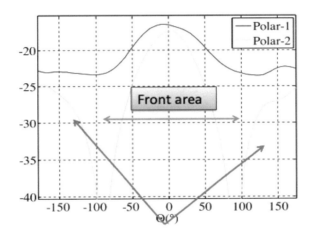

Blades Direction

Figure 4.   Gain array at 900 MHz.

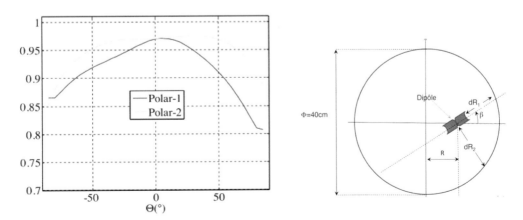

Figure 5.   Depolarization at 900 MHz.

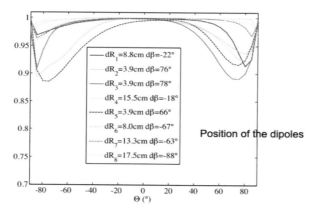

Figure 6.   The array similarity for two polarizations at 900 MHz.

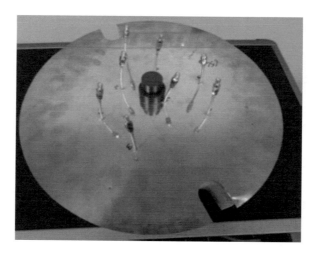

Figure 7.   The ground plane of the array N° 3 with the 8 attenuators.

We note that there is little dispersion on the behavior of reflection coefficients.

Figure 9 presents the photos of arrays 1(a) and 3(b).

The calibration is a very important phase in the characterization of the array. We use this calibration in the calculation of the goniometry. We compare the measured signal of the mobile with the calibration table and with an adequate algorithm the direction of this mobile phone in two different polarizations.

The maximum measured gain is −8 dBi at 700 MHz and 5.9 dBi at 2400 MHz. The results obtained correspond well to the simulations and the goniometry performances have the desired precision on all the frequency bands.

The calibration is a very important phase in the characterization of the array. We use this calibration in the calculation of the goniometry. We compare the measured signal of mobile with the calibration table and with an adequate algorithm the direction of this mobile phone in two different polarizations.

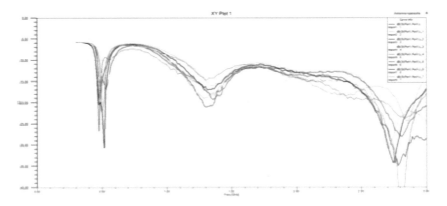

Figure 8. The refection coefficient of the 8th antenna array N° 3.

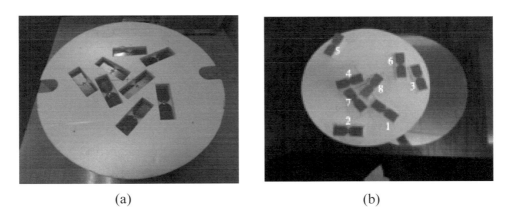

(a)   (b)

Figure 9. Photos of the array 1 (a) and 3 (b).

Figure 10. The calibration of the array N° 3 in the StarLab of the lab.

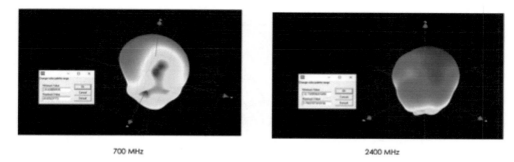

700 MHz   2400 MHz

Figure 11.   The radiation pattern of the antenna 8 at 700 and 2400 MHz of the array N° 1.

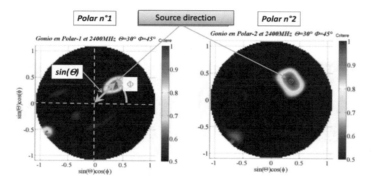

Figure 12.   Results from simulation for goniometry at 2400 MHz.

Figure 13.   Photo of the drone with the array and the receptor ready to be used.

3   CONCLUSION

We designed and optimized an array with 8 identical antenna dipoles operating from 400 to 2400 MHz for the surveillance of disaster areas and people with drones. A dipole was chosen as the basic antenna for its wide bandwidth and stable radiation properties. The dipoles are oriented in a non-uniform manner in order to obtain diversity in polarization. The array performances of the 3 fabricated arrays are validated experimentally.

ACKNOWLEDGMENT

We want to thank BPI France for the financial support for this project SOSPEDRO (SOSPEDRO Project, 2018)

REFERENCES

I. Ben Trad, JM Floc'h, A Ferreol, P Shaley, "Conception d'un Réseau d'Antennes Dipôles pour la Radiolocalisation de Téléphones Portables à Partir de Drones ", JNM 2017, St Malo, 16–19 mai 2017.

P. Cherntanomwong, J. Takada, H. Tsuji and D. Gray, "New Radio Source Localization using Array Antennas based on Fingerprinting Techniques in Outdoor Environment," Asia-Pacific Microwave Conference, 2007.

JM Floc'h, A Ferreol, I Ben Trad, P Thaly, "Design of Dipole Antenna Array for the Radiolocation of Mobile Phones from Drones ", EUC 2018, Munich, 4–6 June 2018

J. Liu, "*Compact Dual-Broadband Antenna and Its Array for 2G/3G/LTE Application*", International Symposium on Computer, Consumer and Control (IS3C), pp 224–226, 2016.

M. G. Pralon; A. Popugaev; D. Schulz; M. A. Hein; R. S. Thomä, "A dual-band compact L-Quad antenna array for radio localization," 9th European Conference on Antennas and Propagation (EuCAP), 2015.

SOSPEDRO Project, final report, September 2018.

B. Stec, M. Czyzewski and A. Slowik, "*Broadband Dipole Antenna Array for S-band*", 15th International Radar Symposium (IRS), 2014.

*Innovative and Intelligent Technology-Based Services for Smart
Environments – Ben Slama et al (eds)
© 2021 Taylor & Francis Group, London, ISBN 978-1-032-02030-3*

Coding metasurface for radar cross section reduction

Akram Boubakri, Saber Dakhli & Fethi Choubani
University of Carthage, SUPCOM, LR11TIC03 Innov'Com Laboratory, Ariana, Tunisia

Tan Hoa Vuong & Jacques David
Plasma and Energy Conversion Lab INPT, France

ABSTRACT: Metasurfaces have recently been known an important interest in manipulating the electromagnetic wave by enabling the control of amplitude, polarization and phase of the incident wave at subwavelength scale. One of the applications of metasurfacing is the Radar Cross Section reduction, widely demanded in military stealth applications by diffracting the incident wave in different directions rather than the radar direction. In this work we propose coding metasurfaces that are basically composed of two binaries (0 and 1) unit cells with respectively 0 and π phase responses. By choosing a different sequence of unit cells, we have been able to control the response of the proposed structure to an incident plane wave with different incidence angles and a significant decrease was observed in both monostatic and bistatic Radar Cross Sections. The studied structures can be used in military applications as the coating of metallic targets to enhance their stealthy features.

1 INTRODUCTION

Recently, several researches have been conducted with the aim to control the propagation of electromagnetic waves. Those studies revealed that metamaterials have a strong ability to surpass the capabilities of natural materials thanks to their unconventional properties and fascinating effects. The key features of them is the resonant behavior controlled by the geometry of the unit cells that are periodically patterned in a subwavelength scale. Based on that, a wide range of applications have been implemented such as metalenses [1,2,3], invisibility cloaking [4] and scattering reduction [5].

Metasurfaces that are considered as the two-dimensional variant of metamaterials with a subwavelength thickness [6] have some mechanical advantages over those three-dimensional bulky structures which are the ease of fabrication and the smaller required volume. Nevertheless, they present uniques feature to manipulate waves regardless their incident angle and polarization from micro-wave to terahertz frequencies.

In general, the metasurface's unit cells are designed to control Surface waves by guiding or splitting them in certain directions through the manipulation of the wave phase at the reflection which is done by controlling the sizes and shapes of the unit cells [7].

As per the effective medium theory, metamaterials and metasurfaces were being classified as "analog metamaterials" and were usually characterized by effective permittivity and permeability [8]. From the perspective of information science, metamaterials and metasurfaces were also represented by digital coding particles '0' and '1', resulting in information metamaterials. Although the resulting metamaterial bytes are still described by the effective medium parameters, the Authors in [9] succeeded in building a digital metamaterial based on some particles that have distinct material properties like positive and negative permittivity.

In addition, one of the most promising applications of metamaterials is the EM stealth [10,11,12], where Radar Cross Section (RCS) reduction remains the principal consideration in the design of military platforms. To the knowledge of the author, metamaterials were used in two different ways

DOI 10.1201/9781003181545-17

for RCS reduction which are the cloaking devices and metamaterial absorbers. For the cloaking shells, the EM waves are guided around the internal objects without perturbing the exterior fields. Nevertheless, the metamaterial absorber, which is much thinner than traditional absorbing materials, utilizes the lossy features of the effective medium parameters to absorb the energy of incident waves.

In this work, we propose a coding metamaterial which is able to manipulate EM waves differently from the two technics mentioned above. By using different coding sequences of '0' and '1' metamaterial unit cells, the proposed structures are able to reduce the monostatic and bistatic RCS over the whole X band regard-less the incident angles and they can be used as a metallic coat for military applications.

2 CODING METASURFACE

2.1 Radar Cross Section

The Radar Cross Section $\sigma(\theta_u, \phi_u)$ of a structure in the direction (θ_u, ϕ_u) is generally defined as [13]:

$$\sigma(\theta_u, \phi_u) = \lim_{r \to \infty} 4\pi r^2 \frac{|E^s(\theta_u, \phi_u)|^2}{|E^i(\theta_i, \phi_i)|^2} \qquad (1)$$

Where E^i (θ_i, ϕ_i) is the incident field due to a plane wave with constant amplitude vector E_0 and E^s (θ_u, ϕ_u) is the scattered far-field in direction θ_u, ϕ_u. Note from (1) that the RCS can be considered a function defined on the unit sphere which did not take into consideration the polarization of the incident and scattered fields which are generally chosen according to θ and ϕ directions. The far-field RCS is mainly the object of practical interest where the monostatic RCS is corresponding to the special case where $(\theta_i = \theta_u, \phi_i = \phi_u)$. For the bistatic RCS the current distribution on the target is being computed for each incidence direction.

2.2 Design of the unit cell

The design of the unit cell is made according to the coding metamaterial concept with 1-bit coding only. For this purpose, we consider a metasurface composed of binary coding elements of '0' or '1' that are physically realized by the '0' element as a unit cell with a 0-phase response and the '1' element as a unit cell with a π phase response in another way the responses of the unit cell is simply defined as:

$$\varphi_n = n\pi \text{ with } n = 0 \text{ or } 1 \qquad (2)$$

The full wave simulator CST microwave studio software is used for our work. As shown in Figure 1, the unit cell is copper square printed over an FR-4 dielectric substrate h=1.4 mm with a ground plane in the bottom. The metallic patch has a thickness of 0.036 mm with an outer square of width w and a non-uniformity of L=1 mm and l=0.5 mm. The inner square has a width of l=0.5 mm. When the outer square width **w** is designed as 4.6 and 3.75 mm, the phase difference is approximately 180° in a broad band as depicted in Figure 2. In particular, from 9 GHz to 12 GHz, the phase difference ranges from 150 to 190 (it is exactly 180° at 8 and 11.5 GHz). Hence, we use the patch element with w=4.6 mm as the '0'element and that with w=3.75 mm as the '1' element. Note that the absolute phase response of the '0' element may not be 0 at a specific frequency, but this case does not affect any physics because the phase can be normalized to 0.

2.3 Analyze and simulation of the proposed structures

Based on the above study, we consider a general square metasurface that contains 3×3 equal-sized lattices with dimension D=64 mm in which each lattice is occupied by a sub array of '0' or '1' elements, as shown in Figure 3 below. The distribution of '0' and '1' lattices is as follows: case

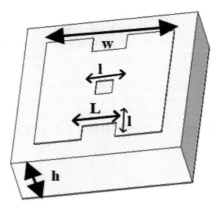

Figure 1.   Caption of the unit cell.

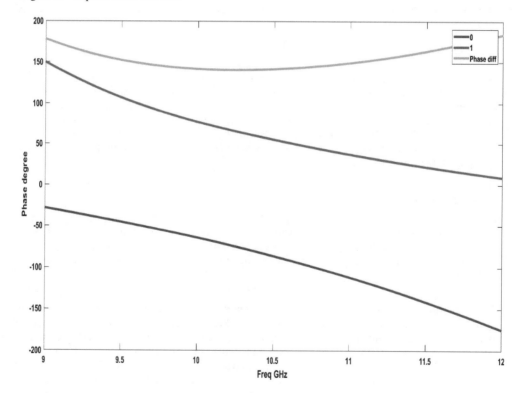

Figure 2.   Caption of the full wave simulation results.

(0101), case 2 (0101/0101) and case 3 (0011/0011). The scattering phase of each lattice is assumed to be ϕ(m,n), which is either 0 or 180°.

   The study of the monostatic RCS for a normal polarized wave (Teta= 0°) showed a remarkable amelioration of the value of RCS compared to a Perfect Electric Conductor (PEC) coated structure for the whole X band as depicted in Figure 4 above.

   The Table 1 compare the simulations results obtained in this work and the achievements of references [14] and [15]. Although the proposed structures in said references showed a 10 dB

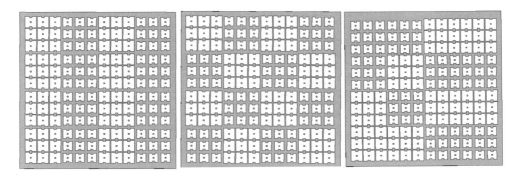

Figure 3. Caption of the three cases of the metasurface square, 0101 (left), 0101/0101 (middle) and 0011/0011 (right).

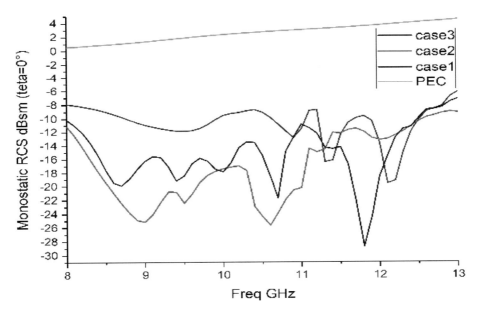

Figure 4. Caption of the monostatic RCS of the proposed structures.

Table 1. Comparison between proposed structures and similar works.

Proposed structure	Coverage of X Band (Monostatic RCS <=10 dB)
[14]	9.4−12 GHz (65%)
[15]	8.38−12 GHz (90%)
This work	8−12 GHz (100%)

reduction of monostatic RCS for a wideband, our proposed structures and particularly case 2 and case 3 are more suitable for counter-measure applications against surveillance RADAR, where a 10 dB RCS reduction is obtained for the whole X Band (8−12 GHz) whereas only 60% and 90% from said band are covered for the reference's structures.

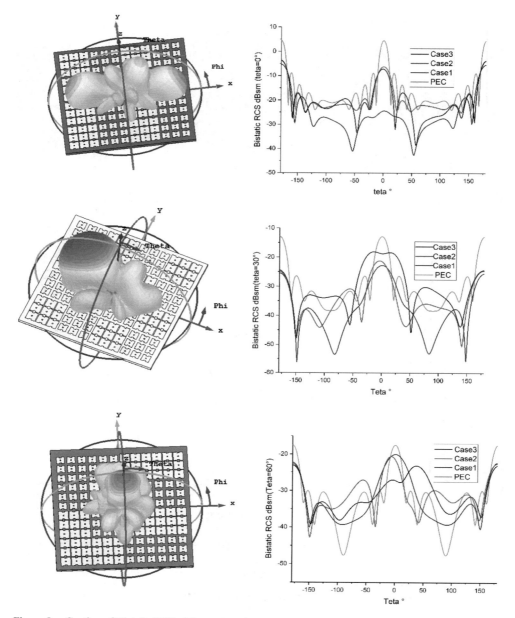

Figure 5.   Caption of Bistatic RCS of the proposed structures at 8 GHz.

As depicted in Figure 5 above, the study of the behavior of the Bistatic RCS with different incident angles for a plane wave θ-polarized, showed an important amelioration at 8 GHz. Case 1 (0101) gives the best Bistatic RCS for the incidence of 0° (about 35 dBsm) and 30°. However, for an incident angle of about 60° the third case (0011/0011) gives the best bistatic RCS which reaches 30 dBsm (10 dBsm enhancement) at the normal direction.

Finally, it's clear that the proposed metasurface structures reflect the incident plane wave into different beam directions as showed in the Figure 4 above which explains with no doubt the enhancement of the bistatic RCS compared to a PEC coated structure.

3 CONCLUSION

In this work we have proposed three types of coding metasurface structures with different sequences of 0 and 1-bits. Compared to a metal plate considered as a reference, a remarkable reduction in the monostatic RCS for the whole X band is realized with a peak of about -30 dBsm for case 3 at around 12 GHz. For bistatic RCS, a peak of about -35 dBsm for case 1 at 8 GHz is obtained and the scattered energy is redirected into more directions for different incident angles of a θ-polarized plane wave. The proposed structures (especially case 2 and 3) are a good candidate for stealth applications as a counter measure against coastal surveillance RADAR usually operating on the X band.

REFERENCES

[1] Pendry JB. 2000. Negative refraction makes a perfect lens. *Phys Rev Lett*, 85, 3966–3969.
[2] Boubakri Akram and all. 2017. A near zero refractive index metalens to focus electromagnetic waves with phase compensation metasurface. *Optical Materials* 69, 432–436.
[3] Boubakri Akram and all. 2014. Subdiffraction resolution super lens based on negative refractive index transmission line metamaterial for free space imaging. *2014 International Conference on Multimedia Computing and Systems (ICMCS) IEEE*, 1367–1371.
[4] Quanlong and all: Efficient flat metasurfaces lens for terahertz imaging. 2014. *Optic express*, 22(21), 25931–25939.
[5] Lai Yun and all. 2009. Illusion optics: the optical transformation of an object into another object. *Physical review letters*, 102 (25), 253902.
[6] Li Aobo and all. 2018. Metasurfaces and their applications. *Nanophotonics*, 7 (6), 989–1011.
[7] Yang and all. 2010. Diffuse reflections by randomly gradient index metamaterials. *Optics letters* 35 (6), 808–810.
[8] Koschny and all. 2004 Effective medium theory of left-handed materials. *Phys Rev Lett*, 93 (10), 107402.
[9] Della Giovampaola and all. 2014. Digital metamaterials. *Nature materials*, 13 (12), 1115–1121.
[10] DE COS and all. 2010. A novel approach for RCS reduction using a combination of artificial magnetic conductors. *Progress In Electromagnetics Research*, 107, 147–159.
[11] Feng, Maochang, and al. 2018. Two-dimensional coding phase gradient metasurface for RCS reduction. *Journal of Physics D: Applied Physics* 51(37), 375103.
[12] Su, Jianxun, and al. 2018. A novel checkerboard metasurface based on optimized multielement phase cancellation for superwideband RCS reduction. *IEEE Transactions on Antennas and Propagation* 66(12), 7091–7099.
[13] KNOTT Eugene. F and all. 2004. Radar cross section. *2nd ed. Sci Tech Publishing*.
[14] Esmaeli, S. H., and S. H. Sedighy. 2015. Wideband radar cross-section reduction by AMC. *Electronics Letters* 52.1: 70–71.
[15] Zhang, Hui and al. 2017. Coding diffusion metasurface for ultra-wideband RCS reduction. *Electronics letters* 53(3), 187–189.

Section II: Smart sensing and artificial intelligence

Innovative and Intelligent Technology-Based Services for Smart Environments – Ben Slama et al (eds)
© 2021 Taylor & Francis Group, London, ISBN 978-1-032-02030-3

A modified speech denoising algorithm based on the continuous wavelet transformer and wiener filter

S. Saoud, M. Bennasr & A. Cherif
ATSEE laboratory, Sciences Faculty of Tunis, University of Tunis El-Manar, Tunis, Tunisia

ABSTRACT: In this work, a speech enhancement interface will be developed in order to be utilized for coding applications, synthesis and recognition. So, we suggest a single channel noise reduction in a noisy environment. The proposed approach combines the Continuous wavelet transform and the wiener filters with the aim of improving the speech intelligibility and quality. The proposed algorithm is tested in various scenarios and shows its robustness, in different noisy environments in term of SNR (signal-to-noise ratio), SSNR (segmental SNR), MSE (Mean Square Error), PESQ (perceptual evaluation of speech quality), Time Evolution and Spectrogram Analysis.

Keywords: Speech denoising; Single channel; Continuous Wavelet Transform; Wiener filter.

1   INTRODUCTION

In recent years, speech enhancement has become an essential stage due to the increased improvement of applications and new technological devices. The goal of speech enhancement is to remove noise from speech signals and to retrieve the original quality of signal that leads the development of the speech intelligibility. In fact, over the last few decades, speech denoising was a challenge field in speech processing.

There has been much research in speech enhancement yet there is still demand for new developments. Various speech denoising algorithms can be categorized into several broad classes: statistical-model-based algorithms, spectral subtractive algorithms, subspace algorithms and wavelet transformers. The first algorithm used for speech enhancement is Spectral subtraction (SS) (Boll 1979; Hu et al. 2002; Lu & Loizou 2008; Cadore et al. 2013) which is easy to implement and supplies a relationship between the remaining noise and signal alteration to a certain extent, but suffers from musical noise which has an abnormal structure and is perceptually boring. The second speech enhancement method is the Wiener filter which is an efficient algorithm commonly used by researchers and is used in many technical applications (Huang et al. 2012). The wiener filter algorithm estimates the optimal noise reduction filter by employing the speech signal and the spectral characteristics of noise. Indeed, the noisy speech is processed via a FIR filter (Finite Impulse Response) whose coefficients are evaluated by reducing the MSE (Mean Square Error) between the clean speech and its estimation to re-establish the enhanced signal. In fact, the Wiener filtering can affect some deterioration on signal. Particularly, for noisy signals at lower levels of SNR, employing the wiener filter algorithm can deteriorate the quality of speech. Therefore, in Wiener filtering methods, the noise reduction amount is in general proportional to speech deteriorations.

In the methods based on the timescale, the noisy signal is at first subdivided into various bands of frequency and the sub signals of reduced noise are after utilized to reconstruct the denoised signal. Indeed, the wavelet transformer is the most effective transformer that can be employed for this subdivision. Many searchers have used the wavelet method and reached some significant results such as the BWT (Bionic Wavelet Transform) (Yao & Zhang 2000). The BWT is an adaptive wavelet transformer founded on the non-linear auditory design of the human cochlear that catches the non-linearity features of the basilar membrane and converts them into adaptive timescale transformations of the appropriate fundamental mother wavelet (Yao & Zhang 2001). The noise reduction-based method is thresholding the adapted BWT coefficients.

The proposed work is inspired from Yao & Zhang 2002; Swami et al. 2015; Sharma & Swami 2014; and Singh & Pritamdas 2016. A novel speech denoising approach based on the continuous wavelet transform and Wiener filter was developed. At first, we employed a discretized continuous wavelet transformer (CWT), which incorporates the human auditory system model into the wavelet transformer (Yuan 2003). Second, the wiener filter is used to enhance the noisy continuous wavelet coefficients. In fact, the suggested algorithm is correctly used to decrease the white and colored noise from noisy speech signal. The results of our algorithm with the TIMIT database (Garofolo 1993) are compared with Wiener filtering and Spectral Subtraction approach.

The remaining sections of this paper are organized as follows: The continuous wavelet transform is detailed in Section 2. In Section 3, we deal with the proposed noise reduction approach. Section 4 presents the experimental results. Conclusion and perspectives are exhibited in Section 5.

2 CONTINUOUS WAVELET TRANSFORM

Morlet is the first who presented the idea of wavelets as a family of functions built from translations and wavelets of a unique function named the "mother wavelet." The wavelet analysis has been presented as a windowing technique with the size of regions being variable. The wavelet decomposition introduces the scale as an alternative to frequency and maps a signal into the timescale plan. The wavelet analysis decomposes the signal into different frequency sine waves (Swami et al. 2015). The continuous wavelet transformer is expressed as Equation 1 below:

$$CWT(a, b) = \frac{1}{\sqrt{a}} \int_{-\infty}^{+\infty} x(t) \cdot \Psi(\frac{t-b}{a}) dt \qquad (1)$$

Where, a = the scale factor of the function; b = the temporal translation of the function; $\Psi(t)$ = the mother wavelet and $\Psi_{a,b}(t)$ = daughter wavelets which are derived from the scale and the shift of the mother wavelet.

For the CWT calculation, the Morlet wavelet was used with a base fundamental frequency of f0 of the unscaled mother wavelet of 15165.4 Hz of the human auditory system. The frequencies were spaced logarithmically in the requested frequency range where the discretized scale variable can be calculated. So, at each scale, the center frequency of the logarithmic spacing is expressed as the following Equation 2:

$$fm = f0/(1.1623)^m, m = 0, 1, 2, \ldots \qquad (2)$$

3 THE PROPOSED METHOD

The proposed work suggests the discretized continuous wavelet transformer (CWT) coefficients and Wiener filtering and afterward calculates the inverse continuous wavelet transformer of the filtered coefficients. The CWT method was employed by Yao and Zhang's original study (Yao & Zhang 2001; 2002). The Morlet wavelet doesn't have scaling functions and thus doesn't support the fast DWT. The support length of the Morlet wavelet is selected as [−4, 4] with different center frequencies for every scale which is given by Equation 2.

In this study, the CWT coefficients are computed at 22 scales (for m = 9–30). The center frequencies corresponding to these scales vary from 166 Hz to 3900 Hz which are logarithmically spaced. We obtain the discretized CWT coefficients by convolving the noisy speech signal with the

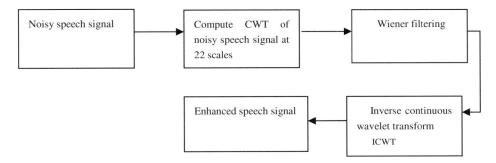

Figure 1.   Block diagram of the proposed method.

scale of the Morlet wavelet to the mentioned frequencies above. The generated CWT coefficients at the requisite number of scales were filtered after via the Wiener filter and reconstructed by the inverse continuous wavelet transform (ICWT). The overview of the proposed approach is given in Figure 1.

4   TEST AND RESULTS

This section presents the experimental results of the proposed noise reduction approach in various noisy environments. For this purpose, in all tests, we have used the speech signal extracted from the TIMIT Acoustic-Phonetic Continuous Speech Corpus (Garofolo 1993) sampled at 16 kHz and recorded per female voice and a various additives noises extracted from Noisex-92 database (Volvo car noise, F16 cockpit noise, factory noise and fhite noise) and at-5 dB, 0 dB, 5 dB and 10 dB were used to evaluate the denoising algorithm.

To examine the performance of the proposed speech enhancement, we have conducted SNR, Segmental SNR, MSE, PESQ (Recommendation, I. T. 2001), time domain waveforms and spectrograms as the criteria of evaluation. The SNR is expressed as follows:

$$SNR = 10 \cdot \log_{10} \left[\frac{\sum_n s^2(n)}{\sum_n s(n) - \tilde{s}(n)^2} \right] \tag{3}$$

Where s (n) = the clean signal and sˆ (n) = the reconstructed signal.

The segmental SNR is defined as:

$$SSNR = \frac{1}{N} \sum_{K=1}^{N} 10 \log_{10} \left[\frac{\sum_{n \in frame_k} |s(n)|^2}{\sum_{n \in frame_k} \left| s^\wedge(n) - s(n) \right|^2} \right] \tag{4}$$

The Mean Square Error (MSE) is expressed as:

$$MSE = \frac{1}{N} \Sigma (si - s^\wedge i)^2 \tag{5}$$

The PESQ (Perceptual Evaluation of Speech Quality) (Recommendation, I. T. 2001) is a measure of objective quality that is agreed upon as ITU-T recommendation P.862. It is developed to estimate the MOS (Mean Opinion Score) subjective test results. The PESQ was conceived to model subjective tests frequently to evaluate voice quality by human beings.

The results of the proposed method shown in Figures 2–6 are compared by the spectral subtraction (SS) and Wiener filter.

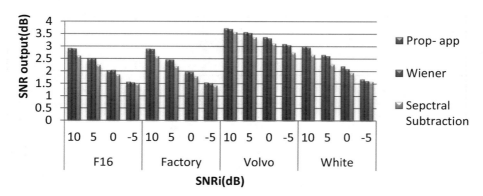

Figure 2. Comparison in terms of SNR.

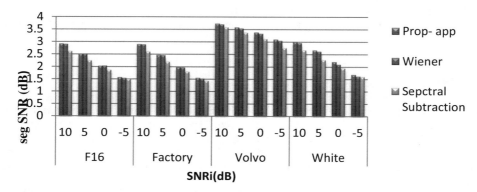

Figure 3. Comparison in terms of Seg SNR.

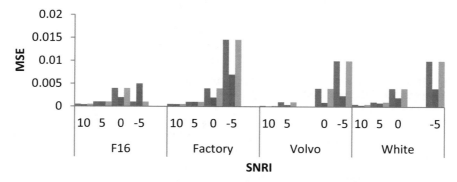

Figure 4. Comparison in terms of MSE.

It's clear from Figure 2 below, that our proposed denoising approach improves the Output SNR more than other methods.

Figure 3 shows the Segmental SNR (seg SNR) results. We observed a significant amelioration in the obtained results especially at −5 dB for different kinds of noises.

It's clear from Figure 4, that the MSE values are less for the proposed approach compared to the wiener filter and the spectral subtraction for various kinds of noise at different SNR levels.

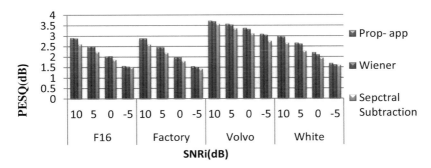

Figure 5.   Comparison in terms of PESQ.

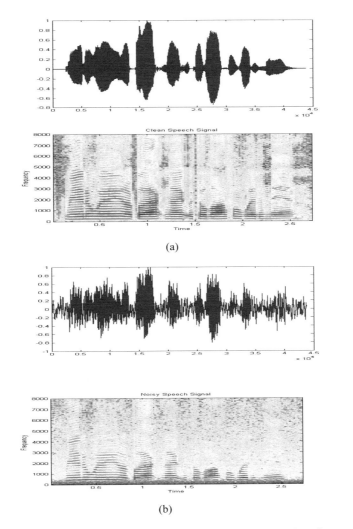

(a)

(b)

Figure 6.   The time evolution and spectrograms results for (a) clean speech; (b) noisy speech altered by factory noise at SNR = 0 dB; (c) denoised speech by Spectral Subtraction; (d) denoised speech by Weiner filter, (e) denoised speech by the proposed approach (SNR = 12.45 dB).

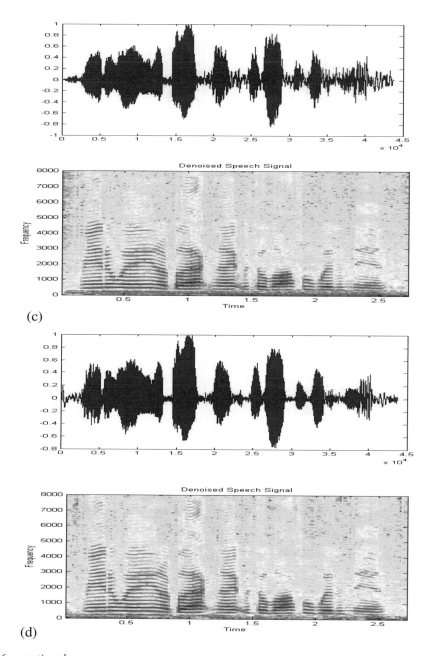

(c)

(d)

Figure 6. *continued.*

Figure 5 shows the PESQ scores for the enhanced speech signal when the speech is degraded by various noises (F16, factory, pink and white).

The PESQ estimation maps the MOS (Mean Opinion Score) and predicts a range between 1 (bad) and 5 (excellent: distortion-less). As clearly shown in Figure 5, the proposed speech enhancement algorithm has the highest PESQ scores, notably, when the speech is corrupted by a VOLVO noise at 0 dB, showing that our method is characterized by the best perceived quality in most cases.

124

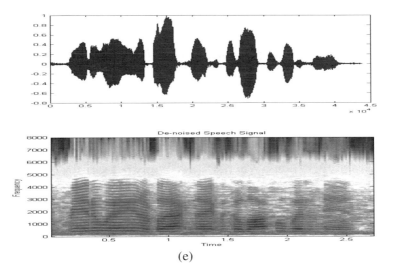

(e)

Figure 6. *continued.*

Figure 6 depicts respectively the time domain waveforms and spectrograms of clean signal, noisy signal and enhanced signals by various methods when the speech is corrupted by factory noise. We observe that the proposed method reduces noise from the speech signal more than spectral subtraction and wiener filter.

5 CONCLUSION

This paper presented a speech denoising approach based on the Continuous Wavelet Transformer using the Morlet Wavelet and the Wiener filter. At first, the coefficients of CWT were scaled perceptually, in relation to the human auditory system and utilized logarithmic scales. Second, the wiener filter is applied to CWT coefficients to remove the noise. Then, by applying the inverse continuous wavelet transformer, we obtained the denoised speech. The comparison of the proposed approach to existing approaches was effectuated by SNR, SSNR, MSE and PESQ. Algorithms were evaluated on speech signals altered by F16 cockpit noise, factory noise, pink noise, Volvo noise and white noise at different input SNR values ranging from −5 dB to 15 dB. The simulation results show the effectiveness of the suggested approach.

The present study is restricted to single channel speech enhancement and needs to be evaluated on multi-channel noise reduction as well as its implementation for real time applications in future work.

REFERENCES

Boll, S. 1979. Suppression of acoustic noise in speech using spectral subtraction. *IEEE Transactions on acoustics, speech, and signal processing* 27(2): 113–120.

Cadore, J., Valverde-Albacete, F. J., Gallardo-Antolín, A., & Peláez-Moreno, C. 2013. Auditory-inspired morphological processing of speech spectrograms. *Applications in automatic speech recognition and speech enhancement. Cognitive computation* 5(4): 426–441.

Garofolo, J. S. 1993. TIMIT acoustic phonetic continuous speech corpus. *Linguistic Data Consortium* 1993.

Hu, H. T., Kuo, F. J., & Wang, H. J. 2002. Supplementary schemes to spectral subtraction for speech enhancement. *Speech Communication* 36(3–4): 205–218.

Huang, F., Lee, T., & Kleijn, W. B. 2012. Transform-domain Wiener filter for speech periodicity enhancement. In *2012 IEEE International Conference on Acoustics, Speech and Signal Processing (ICASSP)*: pp. 4577–4580. IEEE.

Lu, Y., & Loizou, P. C. 2008. A geometric approach to spectral subtraction. *Speech communication* 50(6): 453–466.

Recommendation, I. T. 2001. Perceptual evaluation of speech quality (PESQ): An objective method for end-to-end speech quality assessment of narrow-band telephone networks and speech codecs. *Rec. ITU-T: P. 862.*

Sharma, R., & Swami, P. D. 2014. Enhancement of speech signal by adaptation of scales and thresholds of bionic wavelet transform coefficients. In *Proceedings of: International Conference on Signal Processing, Embedded System and Communication Technologies and their Applications for Sustainable and Renewable Energy (ICSECSRE'14), AVIT,* Paiyanoor, Tamil Nadu, India. ISSN (Print): pp. 2320–3765.

Singh, R.A., &Pritamdas, K. 2016. A Modified Speech Enhancement Algorithm Based on the Continuous Wavelet Transform. *International Journal of Innovative Research in Science, Engineering and Technology* Vol. 5(Issue 5).

Swami, P. D., Sharma, R., Jain, A., & Swami, D. K. 2015. Speech enhancement by noise driven adaptation of perceptual scales and thresholds of continuous wavelet transform coefficients. *Speech Communication* 70 : 1–12.

Yao, J., & Zhang, Y. T. 2000. From otoacoustic emission modeling to bionic wavelet transforms. In *Proceedings of the 22nd Annual International Conference of the IEEE Engineering in Medicine and Biology Society (Cat. No. 00CH37143)* (Vol. 1): pp. 314–316. IEEE.

Yao, J., & Zhang, Y. T. 2001. Bionic wavelet transform: a new time-frequency method based on an auditory model. *IEEE transactions on biomedical engineering* 48(8) : 856–863.

Yao, J., & Zhang, Y. T. 2002. The application of bionic wavelet transform to speech signal processing in cochlear implants using neural network simulations. *IEEE Transactions on Biomedical Engineering* 49(11): 1299–1309.

Yuan, X. 2003. *Auditory Model-based Bionic Wavelet Transform for Speech Enhancement.* M.Sc. Thesis, Marquette University, Speech and Signal Processing Lab Milwaukee, Wisconsin.

Innovative and Intelligent Technology-Based Services for Smart
Environments – Ben Slama et al (eds)
© 2021 Taylor & Francis Group, London, ISBN 978-1-032-02030-3

Analyzing students' digital behavior in an e-learning environment within the blackboard learning management system

A. Bessadok
E-learning and Distance Education Deanship, King AbdulAziz University, Kingdom of Saudi Arabia

E. Abouzinadah & O. Rabie
Faculty of Computing and Information Technology, King AbdulAziz University, Kingdom of Saudi Arabia

ABSTRACT: E-learning has become ubiquitous in many higher education institutions. Learning management systems that deliver E-learning courses generate huge logs of data from student activities. In this paper, we investigated the presence of different student learning activities based on log data. This diversity of students' involvement in the learning process establish through the Learning Management System (LMS) were clustered and the correlation between the e-learning activities was identified. Then, we examined the student specific profiles based on the analyzed data. The study was conducted at King Abdul-Aziz university using the log data obtained from various blended learning courses hosted on the Blackboard platform. The research results showed the existence of three classes defined the profile of each group of students. The correlation test performed was significant and demonstrated the effect of the student's behavioral numerical profile on his/her academic performance.

1 INTRODUCTION

Different students have different behaviors in handling the learning pedagogies. LMS (in this case "Blackboard") can supply educators with information about the way each student interacted with the course's pedagogy. The premise of these systems is teaching each student in a suitable way that fits the learning behavior of each student. Therefore, instructors are interested in knowing their students' learning behavior to be able to deliver the best educational experience. Using artificial intelligence (AI) techniques helps in enriching the educational experience of students (Eck 2007).

Nowadays, one of the main objectives of the education system is to improve learning performance and student satisfaction by using all the necessary and available technological means (Dunn & Kennedy 2019). For this reason, many higher education institutions have initiated the implementation of online learning environment apps for delivering online education resources in a mixed academic learning environment. The importance of the role played by the implementation of E-learning platforms in many higher education institutions is to improve knowledge acquisition, learning perception, satisfaction and motivation, which has been manifested in the exponential growth of the use of LMS (Aldiab et al. 2019; Elfeky et al. 2020). One of the most important goals of e-learning in higher education is to develop a student-centered strategy that can be achieved by implementing a robust LMS platform which guarantees the improvement of the educational process. Therefore, it is crucial to assess and analyze the data generated by the use of LMS. Educational Data Mining (EDM) and Learning Analytics (LA) are considered to be the major fields of research devoted to analyzing data produced from the LMS platform (Aldowah et al. 2019).

2 RELATED WORK

The huge amount of data generated from LMS logs prompted many researchers to deploy appropriate techniques to provide answers with educational interests. Rodrigues et al. 2018 recognized the patterns of student behavior during the learning process. Juhaòák et al. (2019) explored students' behavior and interaction patterns in many types of online quiz-based activities from different courses and with different settings within Moodle learning management systems. They identified types of interaction sequences that shed new light on students' quiz-taking strategies in LMS using process mining methods that took place in three specific phases. Asif et al. (2017) studied two aspects of the performance of undergraduate students using classification and clustering data mining techniques. They identified two groups of students characterized by the worst performing and the best performing students.

Al-Hariri & Al-Hattami (2017) investigated the correlation between student use of technology and their achievements in physiology courses at five health colleges of the University of Dammam. The study involved 231 students studying physiology during their 2nd year in one of the five health colleges (medicine, dentistry, clinical pharmacy, applied medical sciences and nursing) using an online survey. The Pearson correlation coefficient was studied in the correlation between technology and learning achievement in physiology courses. The results showed a significant relationship between use of technology in student achievement. In El-Hmoudova (2015), the researchers conducted a study to examine the learning style of professional English language students. The experiment used Blackboard tests to discover student learning styles by using Index of Learning Styles (ILS) questionnaires. Their results showed that Blackboard can accommodate different learning styles. Lerche & Kiel (2018) used multiple linear regressions of online activity to predict student achievements.

The results showed that online activities can influence learning outcome and success. Abdullah (2015) focused on collecting data for students behaviors in an e-learning environment by using Moodle, from which the data was evaluated using Receiver Operating Characteristics. The evaluation result reflected a correlation between the student performance and the certain learning styles. Hu et al. (2014)'s research shows the effectiveness of analyzing the students' behavior for early warning signs. They used records for 300 students, of which 284 students passed and 16 failed the course. The classifier evaluation reached 97% on warning the students during the class. Mijatovic et al. (2013) used Pearson correlation to study the participation of 169 students using LMS for the first time for their schooling at a Serbian university. The results showed that students participated better in class than by using LMS. In-class participation and interaction with LMS had a significant correlation that may predict the level of learning success. On the other hand, the student Technology Acceptance Model (TAM) does not seem to statistically predict their technology usage, such as using LMS.

3 METHODOLOGY

In this research, we used the k-means clustering method that allowed us to conduct the relationship between student activities. The k-means adds objectivity to the clustering process. Each component of the research method is as follows:

3.1 Data collection

The sample study data was extracted from various e-courses hosted in the platform of Blackboard LMS offered at King Abdul-Aziz University during the first semester of the academic year 2018–2019. The data for this research was based on students' activities on a number of e-courses. Each of these activities had the number of student's hits as shown in Table 1.

Table 1. Sample of the e-courses.

Course	Number of students	Total number of row logs
CPIS 320	59	41.781
CPIS 357	61	54.642
CPIS 444	81	65.435
CPIS 483	72	62.117
Total	301	223.975

3.2 Procedure

The standard retrieved fields in the log files of Blackboard system are characterized as attributes. Table 2 below describes each activity as presented by the accumulator table provided by the Blackboard system.

Table 2. Blackboard Accumulator activities Log dataset attributes*.

Attribute	Description
Pk1	A unique identifier for each record.
Event Type	Groups the event according to what happened.
User pk1	Refers to the Primary Key attribute of the USERS table.
Course pk1	Refers to the Primary Key attribute of the COURSE table.
Group pk1	Refers to the primary key attribute of the GROUP table.
Forum pk1	Refers to the primary key attribute of the FORUM table.
Content pk1	Refers to the Primary Key attribute of the COURSE CONTENT table.
Timestamp	The date and time when the event occurred.
Data	Data linked to the event.
Messages	Additional messages related to the event.
Session ID	Identifies the user session that initiated the action.

*https://help.blackboard.com/Learn/Administrator/Hosting/System_Management/Reports/Running_Database_Attribute_Reports

To characterize the behavior of students through the Blackboard platform, we identified the e-course components as described in the Table 3:

Table 3. Students action identifiers.

Action Identification type	Description
Contents	All files uploaded on the course viewed or edited
Involvement	Activities that have not concerned course materials or assessments
Evaluation	Downloaded quizzes or tests Viewed assignment or finalize homework submission

After several pre-processing steps were applied to the log files provided by the Blackboard Accumulator activity Log dataset attributes, we removed the actions logged by the instructors or guesses as administrators and filtered all non-defined actions.

The remaining rows after the preliminary filtering and cleaning data is 144.214. The attributes of the final dataset are presented in Table 4 below:

Table 4. Final attributes of the complete dataset.

Attribute	Description
Assignment	Number of times the student viewed or took online homework
Course Information	Number of times the student viewed updated information assigned for the course
Getting Started	Number of times the student viewed the content menu of the course
Learning Materials	Number of times the student download or viewed a file
Quizzes/tests	Number of times the student uploaded and submitted a quiz or a test
Student Support	Number of times the student clicked on student support to see if there is any extra content.

4 RESULTS

Clustering is a consistent way of capturing the complexity of student profiles linked to their learning activities. In fact, clustering provides a meaningful explanation of why a particular student is more (or less) active online using the Blackboard learning management system. Several clustering techniques have been deployed to obtain distinct groups with homogeneous user profiles according to the students' activities. The relatively large dataset of the case study example, which can quickly form a cluster based on the log files, can lead to a recommendation of the K-means clustering technique.

The procedure IBM.SPSS 23 consisted of performing several clusters with a different number of clusters chosen a priori until obtaining the optimal number of clusters. The results show the existence of three classes. As presented in the ANOVA table, we can conclude a significant number of clusters obtained at the 0.001 level.

Figure 1 describes the student's intensive path activities. The results indicated that the students in cluster 1 were the most active on the four variables that better described content and involvement but were not interested in frequently consulting the quizzes or in requiring support and are exactly the opposed to the students in cluster 3. The second cluster 2 represents the least active students.

To understand and distinguish the profiles of the students, the figure of the final cluster centers clarifies the commitment of the students for each activity. The profile of students in cluster 1 was characterized by their most significant involvement in the course materials provided, in

Table 5. ANOVA for the standardizing attributes.

	Cluster		Error			
	Mean Square	df	Mean Square	df	F	Sig.
Zscore(Assignments)	10.333	2	.847	122	12.200	.000
Zscore: Course Information	30.302	2	.520	122	58.313	.000
Zscore: Getting Started	38.807	2	.380	122	102.068	.000
Zscore: Learning Materials	30.093	2	.523	122	57.531	.000
Zscore: Quizzes/Tests	23.784	2	.626	122	37.965	.000
Zscore: Student Supports	26.660	2	.579	122	46.017	.000

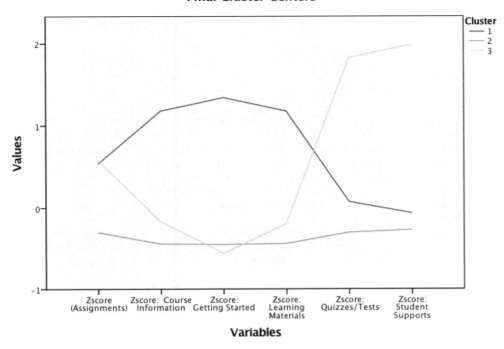

Figure 1. Clustering of students' activities.

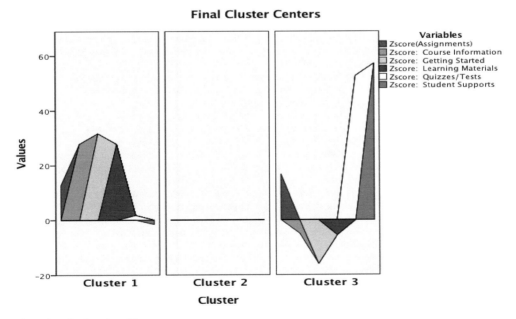

Figure 2. Students' profiles.

exam training and less seeking of supports. The third group describes the profile of the students who feared for the exams and who sought the most help. For the second cluster 2, the students had the average in all the activities, and this was considered as the average profile. (see Figure 2).

5 DISCUSSION AND CONCLUSION

Using artificial intelligence techniques, this study investigated the use of Blackboard and the results showing the ability to classify the students into three groups. This classification can help the instructor in figuring out the types of students they are teaching. The data shows three groups of students; first, students who focus on interacting with the course material and as a result they did not need to ask for help as much as the other groups; second, students who worked more on the assignments and exams, thus, used the help more than the other groups; third, students that did not show focus on materials or assessments and used help more than the first group and less than the second group.

The next step is to use predict the students' behavior via investigating factors as Lerche & Kiel (2018) suggested, e.g., entry qualifications, late applications, disability, age, motivation, living area and tariff points. Another research track is to predict the student groups and their academic success.

REFERENCES

Abdullah, M. A. (2015). Learning Style Classification Based on Student's Behavior in Moodle Learning Management System. *Transactions on Machine Learning and Artificial Intelligence*, *3*(1), 28–28. https://doi.org/10.14738/tmlai.31.868

Aldiab, A., Chowdhury, H., Kootsookos, A., Alam, F., & Allhibi, H. (2019). Utilization of Learning Management Systems (LMSs) in higher education system: A case review for Saudi Arabia. *Energy Procedia*, *160*, 731–737. https://doi.org/10.1016/j.egypro.2019.02.186

Aldowah, H., Al-Samarraie, H., & Fauzy, W. M. (2019). Educational data mining and learning analytics for 21st century higher education: A review and synthesis. *Telematics and Informatics*, *37*, 13–49. https://doi.org/10.1016/j.tele.2019.01.007

Al-Hariri, M. T., & Al-Hattami, A. A. (2017). Impact of students' use of technology on their learning achievements in physiology courses at the University of Dammam. *Journal of Taibah University Medical Sciences*, *12*(1), 82–85. https://doi.org/10.1016/j.jtumed.2016.07.004

Asif, R., Merceron, A., Ali, S. A., & Haider, N. G. (2017). Analyzing undergraduate students' performance using educational data mining. *Computers & Education*, *113*, 177–194. https://doi.org/10.1016/j.compedu.2017.05.007

Dunn, T. J., & Kennedy, M. (2019). Technology Enhanced Learning in higher education; motivations, engagement and academic achievement. *Computers & Education*, *137*, 104–113. https://doi.org/10.1016/j.compedu.2019.04.004

Eck, R. V. (2007). *Building Artificially Intelligent Learning Games* [Chapter]. Games and Simulations in Online Learning: Research and Development Frameworks. https://doi.org/10.4018/978-1-59904-304-3.ch014

Elfeky, A. I. M., Masadeh, T. S. Y., & Elbyaly, M. Y. H. (2020). Advance organizers in flipped classroom via e-learning management system and the promotion of integrated science process skills. *Thinking Skills and Creativity*, *35*, 100622. https://doi.org/10.1016/j.tsc.2019.100622

El-Hmoudova, D. (2015). Innovation of CB achievement professional English language course in Blackboard. *Procedia-Social and Behavioral Sciences*, *171*, 166–171.

Hu, Y.-H., Lo, C.-L., & Shih, S.-P. (2014). Developing early warning systems to predict students' online learning performance. *Computers in Human Behavior*, *36*, 469–478. https://doi.org/10.1016/j.chb.2014.04.002

Juhaňák, L., Zounek, J., & Rohlíková, L. (2019). Using process mining to analyze students' quiz-taking behavior patterns in a learning management system. *Computers in Human Behavior*, *92*, 496–506. https://doi.org/10.1016/j.chb.2017.12.015

Lerche, T., & Kiel, E. (2018). Predicting student achievement in learning management systems by log data analysis. *Computers in Human Behavior*, *89*, 367–372. https://doi.org/10.1016/j.chb.2018.06.015

Mijatovic, I., Cudanov, M., Jednak, S., & Kadijevich, D. M. (2013). How the usage of learning management systems influences student achievement. *Teaching in Higher Education*, *18*(5), 506–517. https://doi.org/10.1080/13562517.2012.753049

Rodrigues, M. W., Isotani, S., & Zárate, L. E. (2018). Educational Data Mining: A review of evaluation process in the e-learning. *Telematics and Informatics*, *35*(6), 1701–1717. https://doi.org/10.1016/j.tele.2018.04.015

Innovative and Intelligent Technology-Based Services for Smart
Environments – Ben Slama et al (eds)
© 2021 Taylor & Francis Group, London, ISBN 978-1-032-02030-3

Deep neural networks for a facial expression recognition system

Nadia Jmour
LA.R.A Laboratory, National Engineering School of Tunis, Tunisia
National Engineering School of Carthage, University of Carthage, Tunisia

Sehla Zayen
LA.R.A Laboratory, National Engineering School of Tunis, Tunisia
Higher Institute of Technological Studies of Charguia, Tunisia

Afef Abdelkrim
LA.R.A Laboratory, National Engineering School of Tunis, Tunisia
National Engineering School of Carthage, University of Carthage, Tunisia

ABSTRACT: Deep learning can be applied to recognizing human facial expression of anger, disgust, fear, happiness, sadness, surprise and neutrality. In this paper, we used the Kaggle (Facial Expression Recognition Challenge), Extended Cohn-Kanade CK+ facial expression datasets and leveraged transfer learning techniques to achieve the best results for emotion classification.

To improve our results, we used VGG-16 and InceptionV3 which were pre-trained on an ImageNet dataset for our convolutional neural networks and we used multi-layer perceptron classifiers as the trained parts. Experiment results show the advantages of using this approach. In addition, the effects of the distribution and the quality of the data for an efficient training and application of this technique is discussed and approved.

1 INTRODUCTION

Recognizing human emotions improves understanding and communication between humans with different situations and opinions by improving our emotional intelligence.

There are several human emotion recognition fields such as facial expression, body posture and voice.

In this paper, we are interested in emotion recognition from facial expressions which is a large field of artificial intelligence. For this purpose, we used facial databases to recognize the 7 emotions identified by Ekman & Friesen 1978: anger, disgust, fear, happiness, sadness and surprise, (Ekman & Friesen 1983).

The first step of the facial expression recognition basic system process is face acquisition. It consists of finding face locations in the input images, (Ebrahimi et al. 2015).

Features extraction is the second step. In this stage, features can be either manually extracted by applying the geometric feature-based methods, the appearance-based methods or learned from training images. The last step is facial expression recognition which uses the extracted features as input for the classifier.

This training process is based on a traditional learning technique to extract features. In this research, we used a deep learning approach and transferred learning to efficiently extract emotion features by leveraging optimized features from the VGG-16 and the InceptionV3 deep neural networks. We also leveraged various datasets such us the Kaggle Facial Expression Recognition Challenge FER-2013 and the Extended Cohn-Kanade CK+ datasets (Lucey et al. 2010).

Our paper is organized as follows: section 2 describes the used databases and the preprocessing steps. Section 3 introduces deep learning architectures and the approaches used in our research. Finally, section 4 is divided into two parts. In the first one, experiment results are presented. In the second part, the effects of the transfer learning technique as well as data quality is discussed.

    DOI 10.1201/9781003181545-20

2 DATA PREPARATION

2.1 *Facial expression databases*

To test our models, we choose the FER-2013 and Extended Cohn-Kanade CK+ datasets. The first one is a wild image dataset and the second one has a tiny number of samples.

2.1.1 *FER-2013 database*

The Kaggle dataset (from the Facial Expression Recognition Challenge) is a wild images dataset, it contains 35,887 images with the format: 48 x 48 pixels (8-bit grayscale). It presents a uniform distribution of data across sex, race and ethnicity. Finally, it contains the seven key emotions.

2.1.2 *CK+ database*

One of the goals we wanted to accomplish is recognizing the seven human expressions.

We needed another database with similar features as those presented by the FER-2013 dataset, but with even greater detail richness and more uniform distribution in terms of subject genders and of classes. CK+ almost perfectly fits these requirements.

This database contains 118 subjects divided into ten groups. It is composed of 327 image sequences. Each sequence starts with a neutral emotion and ends with the peak of the emotion. The peak expression for each sequence is fully FACS coded (Ekman et al. 2002), and given an emotion label. So, we collected the first frame from each sequence for the neutral emotion. In addition, we eliminated the contempt emotion and we select the last frame from each sequence for the first test of our models.

2.1.3 *IBUG 300W database*

The ibug 300W face dataset is built by the Intelligent Behavior Understanding Group (ibug) at Imperial College London. It contains images collected from Google. These images are annotated with 68 facial landmark annotations, (Sagonas et al. 2013).

2.2 *Data preprocessing*

For the VGG-16 model, the input shape has to be (224, 224, 3) or (3, 224, 224). Width and height should be no smaller than 32. Since the format of the FER-2013 images data was: 48 x 48 pixels, we kept the original size and we duplicated the depth to 3 channels.

For the InceptionV3 model, the input shape must be (299, 299, 3) or (3, 299, 299). Width and height should be no smaller than 75.

Making images smaller simply discards pixels. Making them larger adds pixels, which can result in a loss of quality.

For this reason, we resized the original FER-2013 images to 75×75 which is the nearest size to the original 48×48 size and we duplicated the values in 75×75 to three dimensions (48, 48,3).

For the CK+ database, we applied face detection from the original images first. In this case, we resized the images to 224×224 as an input for the two models.

2.3 *Facial expressions tracking*

Landmark initialization and facial tracking consist of detecting important facial structures on the face to localize key points of interest along the image. It is a two steps process:

2.3.1 *Face localization*

Face detection can be achieved by applying OpenCV's built-in Haar cascades detector or a pretrained histogram of oriented gradients and object detection with Linear SVM detectors, or even by using deep learning algorithms for face localization. By applying these detectors, we obtain the (x, y)-coordinates of the facial region so we can obtain face localization, (Dhall et al. 2015).

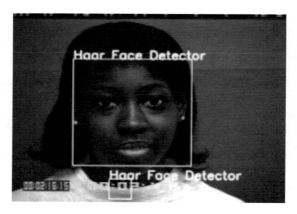

Figure 1.   Haar Cascade face detector.

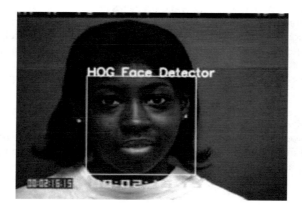

Figure 2.   HOG face detector.

2.3.2   *Detect the key facial structures*

The facial landmark detector included in the dlib library and that we used in our research is a pretrained model proposed by Kazemi & Sullivan in 2014.

This pretrained model is used to estimate the location of 68 (x, y)-coordinates that map facial structures on the face. These 68 point mappings are obtained by a training predictor on the labeled iBUG 300-W dataset, (Sagonas et al. 2013).

In the FER-2013 database, images are collected using Google image search API. To detect faces in every image, researchers have applied OpenCV tools and the incorrectly labeled images are rejected by a human.

For the CK+ database and for every tested input, we applied the Haar Cascade Face Detector in OpenCV (Paul et al. 2001)(Gangopadhyay et al. 2019), and HoG Face Detector in Dlib for face detection on a set of images (Dalal & Trigges 2005). We compared the two methods to choose the one that gives the best predictions on face detection (Kumar & Sharma 2017).

We notice that the first technique gives a lot of false predictions and it doesn't work on non-frontal images. However, the second one works very well for frontal images and gives the right predictions.

Figures 1 and 2 show face detection results for the same image with the two detectors.

Besides that, the HoG face detector is chosen to detect face landmarks in the database. Figure 3 shows an example of a cropped and resized image from CK+ dataset.

Input Image   Face Detection   Cropping   Resizing

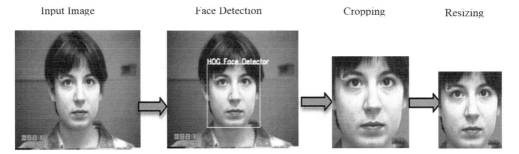

Figure 3.   A CK+ database's image preprocessing.

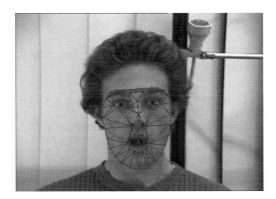

Figure 4.   Face tracking.

2.3.3   *Visualization*

For a better visualization of each of the facial structures and to overlay the results on an input image, the marked feature points are converted to triangles to create a face mask from which key emotion information can be gathered. Figure 4 shows an example of our face tracking application.

3   ARCHITECTURE

In this paper, we aim to recognize human key emotions by using deep learning approaches and transferred learning techniques (Li & Lam 2015). We use the VGG-16 (Li et al. 2019)(Simonyan & Zisserman 2014) and InceptionV3 (Szegedy et al. 2019) deep neural networks.

We choose to work with VGG-16 and InceptionV3 because they achieve efficient results in terms of prediction accuracy and follow a relatively standard CNN architecture. The idea is, first, apply transfer learning by using the learned weights and re-training a few later layers on the FER-2013 and the CK+ facial expression databases and then compare the results for the two deep learning models.

3.1   *VGG-16 deep neural network*

VGG-16 represents one of the most popular convolutional neural network architectures, with 16 layers. It performs 3x3 convolutions and 2x2 pooling. It uses significantly (3×) more parameters than AlexNet, which we have applied in our recent research (Jmour et al. 2018)(Simonyan & Zisserman 2014).

Table 1. FER-2013 and CK+ datasets performance for the VGG-16 and InceptionV3 model.

Model	Databases	Train Accuracy (%)	Test Accuracy (%)	Input shape
VGG-16	Fer2013	98.96	45.83	(48,48,3)
	CK+ (top frame)	100	73.98	(224,224,3)
InceptionV3	Fer2013	97.11	43.22	(75,75,3)
	CK+ (top frame)	99.64	72.35	(224,224,3)

3.2 InceptionV3 deep neural network

The Inception architecture was introduced by (Szegedy et al. 2019).

Inception-V3 is a convolutional neural network trained on more than a million images from the ImageNet database (Deng et al. 2009). It's a deep network with 48 layers.

The Inception model acts as a multi-level feature extractor. It computes 1×1, 3×3 and 5×5 convolutions. The output of these filters is then stacked along the channel dimension before being fed into the next layer in the network.

4 RESULTS AND DISCUSSION

The Pre-processing and training steps are performed using Python libraries such as OpenCV and TensorFlow on GPU NVIDIA version 391.25, a processor type of 16xi5-7300HQ and a memory of 8 GB.

We applied training experiments into 100 epochs, with a learning rate 0.0001 and a 128 for MiniBatchsize in the case of the FER-2013 database and 50 Minibatchsize for the CK+ database by taking into account the effects of this parameter on training performance, (Jmour et al. 2018). We leveraged the Keras implementation of VGG-16 and InceptionV3 (Francois 2015). We then split the CK+ dataset into 90% for learning and 10% for testing.

To evaluate the performance of our models, we looked at the accuracy on the training and test sets and we examined the confusion matrix.

4.1 Effects of transfer learning

This paper investigates transfer learning techniques for the seven key emotions recognition. To accomplish this, we needed information from the related sources to the target domain.

By fine tuning our VGG-16 and the InceptionV3 models using FER-2013, we increase the accuracy by 7% over the baseline method proposed in [18], on the test sets (39.13%). VGG-16 and InceptionV3 had accuracies of 45.83% and 43.22% (Table 1). The biggest advantage of using the VGG16 model was the negligible time to transfer feature vectors and to train the fully connected layers with greater accuracy.

Then, fine-tuning our VGG-16 and the inceptionV3 models using CK+ datasets improved our results and we increased the accuracy to 73.98% and 72.35% respectively (Table 1).

The InceptionV3 model is characterized by using kernels with different sizes in the extraction features part. For global features large kernels are preferred and smaller kernels are used in detecting specific features regions that are distributed across the image frame, (Singh et al. 2020), (Thomas et al. 2020).

The corresponding confusion matrix generated from our model's prediction on the test presented in Figure 5, 6, 7 and 8 clearly improved these results.

4.2 Effects of the data

The overall accuracies along with the CK+ dataset are greater than those on the FER-2013 dataset. VGG-16 achieves its best performance with an accuracy of 73.98% while InceptionV3 achieved

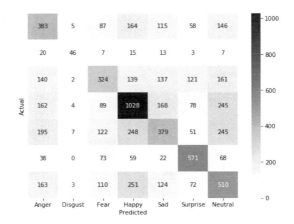

Figure 5. Conf matrix VGG-16 on Fer2013.

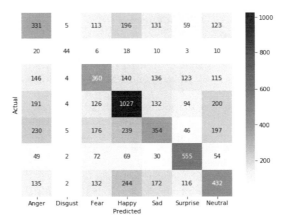

Figure 6. Conf matrix InceptionV3 on Fer2013.

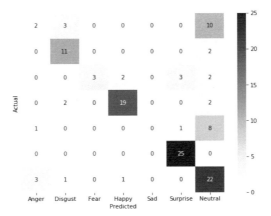

Figure 7. Conf matrix VGG-16 on CK+.

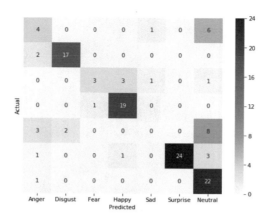

Figure 8. Conf matrix InceptionV3 on CK+.

an accuracy of 72.35% (Table 1). We find it surprising that all models perform better on the CK+ dataset, a significantly smaller dataset than FER-2013.

This may be explained first from the structure and the uniformity of the CK+ dataset in terms of the number of examples and the postures for each subject and each emotion. For example, the happy, the surprise and the neutral classes in the FER-2013 dataset present the highest scoring class, and the classes anger, disgust, sadness and fear are often mistaken for each other by our models.

This can be verified by examining the confusion matrix in Figures 5 and 6.

The disgust class tends to have much fewer training and testing samples compared to the rest of the categories which makes it harder to train the CNN to recognize it.

Deep neural networks need large amounts of data to achieve best performance, but with data where classes are not well distributed or balanced and with millions of images with bad quality, we can't build deep a learning model with high performance.

The InceptionV3 deep neural network goes wider instead of simply going deeper in terms of the number of layers. In addition to quantity, the images in the CK+ dataset are also of higher quality and there are images in the FER-2013 dataset that contain text in the background.

5   CONCLUSION

In this work, we address the task of facial expression recognition and aim to classify
human faces into the seven emotion categories that represent universal human emotions. We use the HoG Face Detector in Dlib for face detection.

We tested the emotion recognition with various techniques, such as fine tuning and we explored different deep neural network architectures. We achieved acceptable results on the Kaggle dataset and we improve our highest accuracy (73.98%) on the CK+ dataset by applying transfer learning. Our results show the importance of this approach on the training as well as the effect of the distribution and the quality of the data used to apply this technique.

ACKNOWLEDGMENTS

This research was supported and financed by Biware Consulting (www.biware-consulting.com). We thank all the groups who provided insight and expertise that greatly assisted the research.

We would like also to express our very great appreciation to M. Amine Boussarsar and M. Walid Kaâbachi for their valuable and constructive suggestions during the planning and development of this research work.

REFERENCES

Dalal, N., Triggs, B.: Histograms of Oriented Gradients for Human Detection. International Conference on Computer Vision & Pattern Recognition (CVPR '05), San Diego, United States. pp. 886–893 (2005).

Deng, J., Dong, W., Socher, R., Li, L.-J., Li, K., and Fei-Fei, L.: ImageNet: A large-scale hierarchical image database. In CVPR, (2009).

Dhall, A., Murthy, R., Goecke, R., Joshi, J., and Gedeon, T.: Video and image based emotion recognition challenges in the wild: Emotiw 2015. In Proceedings of the 17th International Conference on Multimodal Interaction, ICMI '15. ACM (2015).

S. Ebrahimi Kahou, X. Bouthillier, P. Lamblin, C. Gulcehre, V. Michalski, K. Konda, S. Jean, P. Froumenty, Y. Dauphin, N. Boulanger-Lewandowski, R. Chandias Ferrari, M. Mirza, D. Warde-Farley, A. Courville, P. Vincent, R. Memisevic, C. Pal, Y. Bengio, EmoNets: Multimodal deep learning approaches for emotion recognition in video, Journal on Multimodal User Interfaces, (2015).

Ekman, P., Friesen, W.: Facial Action Coding System: A Technique for the Measurement of Facial Movement. Palo Alto: Consulting Psychologists Press (1978).

Ekman, P., Friesen, W.: Emotional facial action coding system. Unpublished manuscript, University of California at San Francisco (1983).

Ekman, P., Friesen, W.V., and Hager, J.C.: Facial Action Coding System: the Manual. Research Nexus, Div. Network Information Research Corp, Salt Lake City, UT (2002).

François, C.: Keras, GitHub repository. https://github.com/fchollet/keras (2015).

Gangopadhyay I., Chatterjee A., Das I.: Face Detection and Expression Recognition Using Haar Cascade Classifier and Fisherface Algorithm. In: Bhattacharyya S., Pal S., Pan I., Das A. (eds) Recent Trends in Signal and Image Processing. Advances in Intelligent Systems and Computing, vol 922. Springer, Singapore (2019).

Jmour, N., Zayen, S., AND Abdelkrim, A. Convolutional neural networks for image classification. In 2018 International Conference on Advanced Systems and Electric Technologies (IC ASET), IEEE, pp. 397–402 (2018).

Kazemi, V., Sullivan, J.: One Millisecond face alignment with an ensemble of regression trees. In: 2014 IEEE Conference on Computer Vision and Pattern Recognition (CVPR), pp. 1867–1874 (2014).

Kumar, Y., Sharma, S.: A systematic survey of facial expression recognition techniques. In 2017 international conference on computing methodologies and communication (ICCMC), pp 1074–1079. ISBN: 978-1-5090-4890-8 (2017).

Lucey, P., Cohn, J.F., Kanade, T., Saragih, J., Ambadar, Z., and Matthews, I.: The extended cohn-kanade dataset (ck+): A complete dataset for action unit and emotion-specified expression. In CVPRW, 2010 IEEE Computer Society Conference on, pages 94–101. IEEE (2010).

Li, J., & Lam, E. Y.: Facial expression recognition using deep neural networks. 2015 IEEE International Conference on Imaging Systems and Techniques (IST). doi:10.1109/ist.2015.7294547 (2015).

Li, K., Jin, Y., Akram, M.W. et al.: Facial expression recognition with convolutional neural networks via a new face cropping and rotation strategy. Vision Computer (2019).

Paul, V., Michael J. Jones: Rapid Object Detection using a Boosted Cascade of Simple Features. Conference on Computer Vision and Pattern Recognition (CVPR), pp. 511–518 (2001).

Sagonas, C., Tzimiropoulos, G., Zafeiriou, S., Pantic, M.: The 300 Faces in-the-Wild Challenge: The First Facial Landmark Localization Challenge. IEEE International Conference on Computer Vision (ICCV) Workshops, pp. 397–403 (2013).

Simonyan K, Zisserman A.: Very deep convolutional networks for large-scale image recognition. ArXiv:1409.1556 (2014).

Singh N.S., Hariharan S., Gupta M.: Facial Recognition Using Deep Learning. In: Jain V., Chaudhary G., Taplamacioglu M., Agarwal M. (eds) Advances in Data Sciences, Security and Applications. Lecture Notes in Electrical Engineering, vol 612. Springer, Singapore (2020).

Szegedy, C. et al.: Going deeper with convolutions, in Proc. IEEE Conference. Computer Vision Pattern Recognition. pp. 1–9 (2015).

Thomas T., P. Vijayaraghavan A., Emmanuel S. (2020) Neural Networks and Face Recognition. In: Machine Learning Approaches in Cyber Security Analytics. Springer, Singapore.

Innovative and Intelligent Technology-Based Services for Smart Environments – Ben Slama et al (eds)
© 2021 Taylor & Francis Group, London, ISBN 978-1-032-02030-3

A reliable data forwarding technique for 5G internet of vehicles networks

Nadjet Azzaoui & Ahmed Korichi
LINATI Lab, Kasdi Merbah University, Ouargla, Algeria
Department of CSIT, Kasdi Merbah University, Ouargla, Algeria

Bouziane Brik
DRIVE Lab, Burgundy University, France

ABSTRACT: Recently, internet of vehicles has become an active field for researchers because it allows efficient and reliable wireless communication between vehicles to everything,;hich has motivated many researchers to develop communication paradigms that meet emerging vehicle requirements in smart cities. Dedicated Short Range Communication-based approaches and cellular technologies-based approaches such as 5G are the most promising technologies for vehicular communications. In order to provide efficient data transmission, seamless and integrated communication between vehicles to everything, this contribution proposes a reliable data forwarding technique for the Internet of Vehicles called DFS-IoV. Our technique is based on 5G Cellular-Vehicles connecting to everything. An extensive evaluation is conducted to evaluate the performance of the proposed technique under various network conditions in urban scenarios. The results of simulation show that the suggested technique ensures increased throughput, enhances the packet delivery ratio and guarantees reliability in comparison to the already existing techniques.

Keywords: Smart_Cities, IoV, V2X, DSRC, 5G C-V2X, Data Forwarding.

1   INTRODUCTION

Recently, to take advantage of the Internet connection everywhere in smart cities, the vehicular Ad Hoc Network (VANET) is being transformed into the Internet of Vehicles (IoV) by Internet of Things (IoT) technology (Azzaoui et al. 2019). In fact, the IoV is a powerful paradigm that integrates vehicles to everything (V2X): vehicles to pedestrians (V2P), vehicles to homes (V2H), vehicles to sensor (V2S) interactions, and in addition to the integration of conventional vehicles to infrastructure (V2I) and vehicle to vehicle (V2V) communication modes (Sherazi et al. 2019). This enables support for more advanced applications such as autonomous vehicles and collision avoidance between them (Xu et al. 2017).

V2X communication allows the exchange of data between vehicles and with other devices via two main technologies: dedicated short-range communications (DSRC) which is based on IEEE 802.11p, and cellular vehicle-to-everything technology (C-V2X) such as 4G/LTE and 5G (Kiela et al. 2020).

DSRC wireless technology is used between vehicles, pedestrians, and roadside infrastructure to exchange data when the distance is limited to hundreds of meters (Kiela et al. 2020). Short-range data exchange is conducted between the devices that are installed in DSRC communication system vehicles, such as onboard units (OBUs), roadside units (RSUs), or pedestrians' portable mobile devices (Abboud et al. 2016).

C-V2X is a cellular communication technology for transferring data between V2X. in reality, C-V2X is a very powerful complement to DSRC technology. It is defined by the Third Generation

DOI 10.1201/9781003181545-21

Partnership Project (3GPP) which includes LTE and 5G based V2X systems (Yang & Hua 2019). Compared to DSRC, C-V2X can keep pace with changes and adapt to more complex security application scenarios to meet low response time and high reliability as well as meeting bandwidth requirements. Effective communication distances can reach more than two times the DSRC communication distance (Yang & Hua 2019).

The transmitted data require low latency, high data rates, and high QoS. The use of 4G to transfer data can reduce many issues. Yet, 4G doesn't meet the needs of these devices (Benalia et al. 2020) owing to the increase of the number of connected devices. Moreover, the heterogeneity of sub-networks that include 4G LTE, DSRC and nodes forming the IoV network make the forwarding of a data packet with high Reliability in real-time from node source to destination more difficult (Khan et al. 2018).

The advent of 5G wireless technology that is based on the IEEE802.11 ac standard (5G) will provide high speeds and greater coverage than the current 4G (Benalia et al. 2020; Khan et al. 2018). Thus, the principal objective from this work is to provide a new idea about the forwarding of data with low exchange time, and high reliable V2X communication between V2X. In the present work, we propose a reliable data forwarding technique for the Internet of Vehicles called "DFS-IoV" based on 5G Cellular-V2X (C-V2X). Thus, each vehicle will be equipped with a 5G wireless interface and positioning device GPS, in addition to the conventional Short-Range data exchange devices that are installed in the vehicle. Therefore, the idea of making use of new Technologies is meant to guarantee the continuity and the reliability of data flow between the connected devices.

An extensive evaluation is conducted to evaluate the performance of the proposed technique in an urban scenario. The result of simulations shows that the suggested technique ensures increased throughput, enhances packet delivery ratios and guarantees reliability in comparison to the already existing techniques.

The remainder of this paper is organized as follows. Section 2 introduces the relevant related work on wireless communication technologies for data exchange in the vehicular networks. In section 3, we describe the proposed technique in IoV. Afterward, in section 4, the evaluation of our technique is discussed. Finally, section 5 concludes the paper.

2 RELATED WORKS

In this section, we provide the most important wireless communication technologies for data transmission in vehicular networks.

Network access technologies of IoV are consider by authors in Yang et al. 2017; Mahmood et al. 2019; and Kaiwartya et al. 2016 to be classified into access technologies that are both inter-vehicle networks, which include DSRC and WAVE, and mobile Internet access which includes LTE and WiMAX-WLAN, on which vehicles rely to access networks and to realize the communication between vehicles and networks. Heterogeneous approaches aiming to incorporate the advantages of both DSRC and LTE are also mentioned in (Mahmood et al. 2019). In Shabir et al. (2019), to ensure the low collision rate and maximum transmission probability of the high priority information, a multi-priority distributed channel congestion control approach based on IEEE 802.11p is proposed. However, authors in Storck et al. (2019) argued that those objectives are barely met through the IEEE 802.11p technology that has a 27 Mbps maximum throughput with V2X applications. Alternatively, LTE technology has been used to back vehicle applications, yet its performance is affected by interference.

According to cellular strategy (Sanchez-Iborra et al. 2017), some problems involve a previously established framework that cannot be obtained in remote scenarios. Some studies (Syfullah & Lim 2017; Tseng et al. 2019; Abada et al. 2020) have suggested an alternative to using 4G technology or combining transfer technologies of a different nature to deploy hybrid systems (V2I and V2V). In contrast, authors (Sanchez-Iborra et al. 2017) believe that the dedicated short-range solution provides a very probable solution.

Table 1. Overall properties of DSRC and C-V2X.

Topic	DSRC 802.11p	C-X2V
Objectives	Direct safety communication in real-time between (V2R & V2I)	
Deployment	Since 2017. Mass marketing 2019	Initial deployments will actually take place in 2021
	Hybrid model. Possible use with any cellular network(4G/5G) for non-safety services	
Cellular connectivity	Self-managed	Cellular operators can optionally implement real-time theoretical control to increase network usage.
Security	OFDM with CSMA offers strong communication in a dense and dynamic environment. It has no dependency in the presce of the GPS signal.	SC-FDM with semi-persistent sensing coding gain thanks to turbo coding and HARQ
Infrastructure Investment	Public key Cryptography and Infrastructure can enhance safety	
Roadmap	802.11p: targets interoperability With 802.11p	C-V2X Rel.16: Based on NR(5G). It operates in a different channel than Rel.14/15

(Bazzi et al. 2019) see that all of the aforementioned works agree that unlike the IEEE 802.11p, C-V2X can provide a longer range. However, most of them support settings that may not be ideal for IEEE 802.11p. Table 1 highlights the similarities and differences between the usability and overall properties of DSRC and C-V2X.

Huang et al., in 2017, introduced SDVNs that support 5G, where SDN is employed to manage neighborhood groups of vehicles in a vehicular environment and 5G. An architectural control network for 5G-SDVN was introduced by means of combining MEC and SDN.

After scrutinizing a set of related works, one concludes that, so far, no single work has proposed the incorporation of the 5G C-V2X in IoVs networks.

3   OUR DATA FORWARDING TECHNIQUE FOR IOV (DFS-IOV)

DFS-IOV is a reliable data forwarding technique which relies on C-V2X communication technologies for IoV. One assumes that each vehicle is equipped with a positioning device GPS and 5G C-V2X wireless interface, as shown in Figure 1.

This technique allows the use of cellular 5G networks and exploits short-range communications between vehicles for data exchange. Specifically, the 5G permits forwarding data packets quickly and reliably.

The data packets being forwarded from the nodes (i.e., vehicles) need to know the position, direction, and route information to forward the data to the destination nodes.

Figure 2 shows the flowchart of the V2X communication model. The vehicle will be constantly monitoring its environment and transmitting data packets via 5G C-V2X/DSRC. It would check if the destination vehicle is in range, bandwidth, etc.

4.1 *Packet delivery ratio*

PDR is an essential metric of routing protocol performance in any network. This metric defines the percentage of packets successfully received at the destination.

Figure 3 shows that the highest packet delivery ratio belongs to the 5G based routing protocol.

For all these protocols, we noticed an instability of PDR when the number of vehicles increases from 40 to 160. On the other hand, in the case of increasing vehicles from 160 to 200, we observed that the PDR increases.

4.2 *Throughput*

Throughput is a metric that defines the average rate of successfully transmitted data packets.

Figure 4 shows that the 5G based routing protocol has better throughput than the other routing protocols based on DSRC and LTE (Hernafi et al. 2017; van Glabbeek et al. 2016) via a changing number of vehicles. Besides that, the graphs of routing protocols based on DSRC, LTE show the throughput instability as the number of vehicles increases. We see a decrease in throughput for all these techniques from 40 to 80 and from 160 to 200. Additionally, changing the number of vehicles from 40 to 200 has a slight decrease in the throughput of the 5G-based routing protocol.

5   CONCLUSION

The present work proposes a reliable data forwarding technique for IoV based on 5G C-V2X communication technologies, which enables V2X communication to support the IoVs network. The initial individual evaluation, based on high PDR and throughput, demonstrates that the proposed FDS-IoV technique outperforms current forwarding techniques which are based on other wireless technologies such as DSRC, 4G/LTE.

As future work, one would investigate the FDS-IoV technique simulations by making use of an advanced ecosystem, evaluating the safety messages exchanged between the vehicles in highway and urban scenarios.

REFERENCES

Abada, D., Massaq, A., Boulouz, A. and Salah, M.B., 2020. An adaptive vehicular relay and gateway selection schemetechnique for connecting VANETs to internet via 4G LTE cellular network. In *Emerging Technologies for Connected Internet of Vehicles and Intelligent Transportation System Networks* (pp. 149–163). Springer, Cham.

Abboud, K., Omar, H.A. and Zhuang, W., 2016. Interworking of DSRC and cellular network technologies for V2X communications: A survey. *IEEE transactions on vehicular technology*, *65*(12), pp. 9457–9470.

Azzaoui, N., Korichi, A., Brik, B., el amine Fekair, M. and Kerrache, C.A., 2019, October. On the Communication Strategies in Heterogeneous Internet of Vehicles. In *The Proceedings of the Third International Conference on Smart City Applications* (pp. 783–795). Springer, Cham.

Bazzi, A., Cecchini, G., Menarini, M., Masini, B.M. and Zanella, A., 2019. Survey and perspectives of vehicular Wi-Fi versus sidelink cellular-V2X in the 5G era. *Future Internet*, *11*(6), p. 122.

Benalia, E., Bitam, S. and Mellouk, A., 2020. Data dissemination for Internet of vehicle based on 5G communications: A survey. *Transactions on Emerging Telecommunications Technologies*, *31*(5), p. e3881.

Hernafi, Y., Ahmed, M.B. and Bouhorma, M., 2017. ACO and PSO algorithms for developing a new communication model for VANET applications in smart cities. *Wireless Personal Communications*, *96*(2), pp. 2039–2075.

Huang, X., Yu, R., Kang, J., He, Y. and Zhang, Y., 2017. Exploring mobile edge computing for 5G-enabled software defined vehicular networks. *IEEE Wireless Communications*, *24*(6), pp. 55–63.

Kaiwartya, O., Abdullah, A.H., Cao, Y., Altameem, A., Prasad, M., Lin, C.T. and Liu, X., 2016. Internet of vehicles: Motivation, layered architecture, network model, challenges, and future aspects. *IEEE Access*, *4*, pp. 5356–5373.

Khan, S.M., Chowdhury, M., Rahman, M. and Islam, M., 2018. Feasibility of 5G mm-wave communication for connected autonomous vehicles. *arXiv preprint arXiv:1808.04517.*

Kiela, K., Barzdenas, V., Jurgo, M., Macaitis, V., Rafanavicius, J., Vasjanov, A., Kladovscikov, L. and Navickas, R., 2020. Review of V2X–IoT Standards and Frameworks for ITS Applications. *Applied Sciences, 10*(12), p. 4314.

Mahmood, A., Zhang, W.E. and Sheng, Q.Z., 2019. Software-defined heterogeneous vehicular networking: The architectural design and open challenges. *Future Internet, 11*(3), p. 70.

Sanchez-Iborra, R., Gómez, J.S., Santa, J., Fernández, P.J. and Skarmeta, A.F., 2017. Integrating LP-WAN communications within the vehicular ecosystem. *J. Internet Serv. Inf. Secur., 7*(4), pp. 45–56.

Shabir, B., Khan, M.A., Rahman, A.U., Malik, A.W. and Wahid, A., 2019. Congestion avoidance in vehicular networks: A contemporary survey. *IEEE Access, 7*, pp. 173196–173215.

Sherazi, H.H.R., Khan, Z.A., Iqbal, R., Rizwan, S., Imran, M.A. and Awan, K., 2019. A heterogeneous IoV architecture for data forwarding in vehicle to infrastructure communication. *Mobile Information Systems, 2019.*

Storck, C.R. and Duarte-Figueiredo, F., 2019. A 5G V2X ecosystem providing Internet of vehicles. *Sensors, 19*(3), p. 550.

Syfullah, M. and Lim, J.M.Y., 2017, February. Data broadcasting on Cloud-VANET for IEEE 802.11 p and LTE hybrid VANET architectures. In *2017 3rd International Conference on Computational Intelligence & Communication Technology (CICT)* (pp. 1–6). IEEE.

Tseng, H.W., Wu, R.Y. and Lo, C.W., 2019. A stable clustering algorithm using the traffic regularity of buses in urban VANET scenarios. *Wireless Networks*, pp. 1–15.

van Glabbeek, R., Höfner, P., Portmann, M. and Tan, W.L., 2016. Modelling and verifying the AODV routing protocol. *Distributed Computing, 29*(4), pp. 279–315.

Xu, Z., Li, X., Zhao, X., Zhang, M.H. and Wang, Z., 2017. DSRC versus 4G-LTE for connected vehicle applications: A study on field experiments of vehicular communication performance. *Journal of Advanced Transportation, 2017.*

Yang, F., Li, J., Lei, T. and Wang, S., 2017. Architecture and key technologies for Internet of Vehicles: a survey.

Yang, Y. and Hua, K., 2019. Emerging technologies for 5G-enabled vehicular networks. *IEEE Access, 7*, pp. 181117–181141.

*Innovative and Intelligent Technology-Based Services for Smart
Environments – Ben Slama et al (eds)*
© 2021 Taylor & Francis Group, London, ISBN 978-1-032-02030-3

FR-VC: A novel approach to finding resources in the vehicular cloud

Mohamed Ben Bezziane, Ahmed Korichi, Mohamed el Amine Fekair & Nadjet Azzaoui
Computer Science and Information Technologies Department, University Kasdi Merbah Ouargla, Algeria

ABSTRACT: The cloud computing concept has been introduced as a solution that helps to share services in a mobile environment. The combination of the concept of cloud computing and an ad-hoc vehicular networks (VANET) forms a Vehicular Cloud (VC), which offers services to consumer vehicles via supplier vehicles. These services have a tremendous effect on internet connectivity, storage and data applications. Due to the high-speed mobility of vehicles, users in consumer vehicles need to find out services in their neighborhoods. Moreover, supplier vehicles offer services with different qualities, which conduct consumer vehicles to select the best of them. In this paper, we propose a novel approach (FR-VC) that allows constructing the VC using the Roadside Unit (RSU) directory and the Cluster Head (CH) directory to give more visibility to supplier vehicle resources. Clusters of vehicles that move on the same road form a mobile cloud. Whereas the other vehicles form a different cloud in the roadside unit. Simulation results prove the effectiveness of our approach in terms of wait time to register and hit percentage metrics.

Keywords: Vehicular cloud, vehicular ad hoc network (VANET), vehicle clustering.

1   INTRODUCTION

In the last decade, we have witnessed the advancement of technologies in vehicular ad hoc networks (VANETs) aiming to enhance driving by improving safety and providing commercial services (Li et al. 2018) to increase the utility of Intelligent Transportation Systems (ITS).

VANTs are deemed a specific type of Mobile Ad-Hoc Network (MANET) with special movement and limited range. The vehicles circulate on the roads according to an organized map termed a mobility model, based on predetermined roads and buildings, etc. (Hasrouny et al. 2017). Two types of VANET nodes can be differentiated:

1. Mobile nodes including vehicles.
2. Fixed nodes that are identified as roadside units (RSUs) which are installed at strategic points on side of the roads.

Vehicles and RSUs forward messages between the source nodes and the destination nodes using wireless and multi-hops of two types of digital data transmission, which are the inter-vehicle (IVC) and vehicle to road (VRC) communications. As a result, drivers, passengers, and transportation authorities benefit for the management of the freeways, safety, road weather and drivers helping (Dutta & Thalore 2017).

Intelligent vehicles in smart cities are supplied with certain sensors: wireless OBU on-board unit and radars, GPS (Global Positioning System) tracking systems, processing power to extend the coverage of each individual recall with the perception of other vehicles. Likewise, RSUs promise high-speed wireless internet access (LTE, WiMAX, and 5G) (Qureshi et al. 2018). However, the systems experienced delays, intermittent connections, and lost packets due to the high mobility of vehicles.

Undeniably, Mobile Cloud Computing (MCC) can comparatively resolve obvious problems that arise with the growing number of connected devices in traditional cloud computing. (Lo'ai et al.

2016), subsequently, the MCC implies the emergence of the vehicular cloud (VC). While some vehicles behave like suppliers in VC, others behave like customers (Brik et al. 2018).

Note that a supplier vehicle can offer its resources to neighboring vehicles. Besides, the latter is carried out according to three basic types of services with their attributes, which are: network as a service (NaaS) to offer additional bandwidth in terms of the internet to others, storage as a service (SaaS) to offer storage, and Data as a service (DaaS) service to offer data such as books, city maps and videos (Mershad & Artail 2013). The main idea behind our proposed approach is to save all services containing their attributes in a repository on the nearest RSU or at the cluster head level, to have more visibility for consumer vehicles that request RSUs or cluster-heads for the availability of services to consume. Therefore, simulation results regarding Wait Time to Register (W_time) and Hit Percentage (HIT) are calculated and discussed against other work.

The rest of the paper is organized as follows: in Section II, we present the related work. Section III is dedicated to our novel approach and its technical unfolding. Section IV shows the simulation that validates our proposal and discusses the reached results. Finally, Section V concludes this presented work.

2 RELATED WORK

This section highlights the existing vehicular clouds proposed for discovering and consuming services. Whereas, several architectures and protocols have been proposed in this context. To properly situate our research towards a specific work, we are interested in all subjects related to protocols that discover and offer the resources on the vehicular cloud.

The authors in Mershad & Artail (2013) introduced discovering and consuming services with in-vehicular cloud protocols CROWN where they regarded an urban area and mentioned a set of vehicles as suppliers known as STARs and the other as consumers. The STAR that offers the services must find the nearest RSU to provide their resources to consumers and propose them. When a user in a smart vehicle wants the internet connection, they require extra resources that his vehicle is not equipped in or is interested in certain data. Hence, he can demand the services of one or more nearby STARs by requesting the RSU. To achieve consumption, the latter searches for STARs in its directory, which contains all information about the resources of STARs and answers consumers.

In Brik et al. 2018, the authors implemented a DCCS-VC Discovering and Consuming Cloud Services in Vehicular Clouds protocol, where they considered the buses as getaways to access the VC for consuming resources. This proposal is based on finding adequate public buses and using them as cloud directories to facilitate the discovery of services' provider vehicles. This protocol was proposed due to the restricted trajectory of buses, which provides the predictability of time and space. When the vehicle needs to consume services has to connect with the nearest bus.

Jafari et al. (Jafari Kaleibar et al. 2020) implemented a Two-level cOntroller-based aPproach for serVice advertISement and discOveRy in vehicular cloud networks (TOPVISOR). In which they proposed an approach as a VC. The proposal uses two central controllers connected between them and with the distributed directories at the RSUs, the discovery process is accomplished by the hierarchical proactive method at the controller levels.

3 PROPOSED APPROACH

We designed the FR-VC approach termed Finding Resources in the Vehicular Cloud considering the MANHATTAN mobility model (Nagel & Eichler 2008) as a map, where all vehicles in that area move randomly. RSUs are chosen near to the junctions.

3.1 VC creation at the cluster head level

The vehicles communicate directly with each other via their OBU antennas or with the nearest RSUs. The RSUs and the routing protocols used by vehicles makes it possible to build a set of clusters if they receive the services offered or the requests according to the vehicle needs of consumers who are on the same trajectory. These RSUs should only establish a cluster when they receive a set of registrations or request packets in its neighborhood or via a routing protocols. Furthermore, to establish a cluster, it must elect a cluster head that will be able to control the cluster and respond to the consumer vehicles' demands which are referring to this cluster. The remainder of the vehicles that do not meet the circumstances mentioned above must register their services in the most adjacent RSU if they are supplier vehicles or send their requests if they are general public vehicles. Nevertheless, with a random delay value (*dly*) between [0.05, 0.1] seconds, all RSUs are linked between them by a wired transmission, which is delayed by this value.

When a supplier vehicle desires to offer its resources to other vehicles, it sends a REGISTRATION PACKET to the closest RSU, which contains:

- The ID of the vehicle.
- The average speed of the vehicle.
- The direction of the vehicle.
- The geographical coordinates of supplier vehicle.
- The time to leave the VANET Tl.
- The offered resources.

Sending a REGISTRATION PACKET to the RSU from the supplier vehicle (its location is R1) uses the routing protocol. We used geography-based routing such as Enhancement of Greedy Perimeter Stateless Routing Protocol (E-GPSR) (Houmer & Hasnaoui 2019). This kind of routing uses the positions to forward packets from the destination node to the source node and this is our case (we use the coordinates of vehicles).

A consumer vehicle that wants to consume certain services can use the services of one or more nearby supplier vehicles by sending a REQUEST PACKET to the closest RSU, which contains:

- The ID of the vehicle.
- The average speed of the vehicle.
- The direction of the vehicle.
- The geographical coordinates of the vehicle.
- The requested resources.

Accordingly, RSU's collection of registration and request packets launches certain processes to establish a cluster and pick a cluster head or build a directory at its level.

Meanwhile, the holding of the first packet is a registration packet or a request packet. The RSU starts a clock, Then the sending process of packets is consecutive until the clock reaches a certain threshold, *Tth,* for example thirty (90) seconds (which we haven't chosen randomly, but it comes down to several simulation attempts), during which some actions are launched.

3.1.1 RSU level actions

This task is called a discovery phase which enables all resources to be saved on the RSU directories or to create cluster head directories; we outlined these activities as follows:

- Reset clock to zero.
- Put all packets on a list (ListV).
- If ListV contains only supplier vehicles, the RSU creates its proper directory (because we do not need to create a cluster that will only contain similar vehicles, and it will have no offering or consumption operations in the cluster), this is the second case of the RSU-based vehicular cloud.

- If ListV contains only consumer vehicles, the RSU creates its proper directory (same reasons as mentioned above). This is the second case of RSU-based vehicular cloud.
- If ListV contains only one supplier vehicle or one consumer vehicle, the RSU creates its proper directory, and this is the second case of RSU-based vehicular cloud.
- If ListV contains at least one supplier vehicle or one consumer vehicle, the RSU creates a cluster and selects its cluster head (CH).
- Delete ListV's items in DR after creating the cluster.

3.1.2 *Cluster creation*

To create a cluster, we defined what is called a cluster section, by dividing all routes on the map into sections allowing building clusters. With this technique, RSU can recognize the vehicles that belong to the same cluster section and proceed to create the cluster. Each cluster section has a length Ls and a width Ws. Vehicles that circulate on the roads have the coordinates in this plan. So, to select a CH_j, we used the Euclidean distance, taking the coordinates the parameters of the vehicles $V_i(X_i, Y_i)$ and their average speeds AV_i, We calculated the means of X_j, Y_j and AV_j by the Equations 1,2 and 3 respectively below:

$$Xj = \frac{\sum_{i=1}^{n} X_i}{n} \tag{1}$$

$$Yj = \frac{\sum_{i=1}^{n} Y_i}{n} \tag{2}$$

$$AVj = \frac{\sum_{i=1}^{n} AV_i}{n} \tag{3}$$

where n is the number of vehicles in cluster section S_j.

Now we have a point $P(X_j, Y_j)$. To obtain the coordinates of the vehicle that acts as a cluster head in the section S$j$taking the average speed as a factor, we calculated the distance $Dist$ between all vehicles and P. The vehicle that has a shorter distance will be a CH_j.

The Euclidean distance between vehicle Vi and P is calculated by this formula in Equation 4.

$$Dist(V_i, P_i) = \sqrt{\left(X_{Vi} - X_j\right)^2 + \left(Y_{Vi} - Y_j\right)^2 + \left(AV_{Vi} - AV_j\right)^2} \tag{4}$$

The RSUs transmit to all vehicles, which will be on the same cluster for this creation by broadcasting the CLUSTER PACKET. The cluster packet contains the ID of the cluster head and all information about vehicles that are belonging to the same cluster.

The role of the CH is to control the operation of consumption between vehicles. This is done to optimally schedule the queues of consumers. Now the CH knows all consumer vehicles, and then it searches in its list that it received from RSU during the creation process. It then returns via a RESPONSE PACKET that holds the ID of supplier vehicles and their resources.

The consumer vehicle designates the needed resources via transmitting a SERVICE PACKET to supplier vehicles using the routing protocol. The service packet includes the identification of the vehicle and its requests, upon receiving of the vehicle service packet, the supplier vehicle answers either with a NEGATIVE ACKNOWLEDGMENT if it cannot meet the user's request or its waiting line is busy or with a SERVICE RESPONSE packet, inviting the consumer vehicle for the method of payment. The user from consumer vehicles and the supplier vehicles can negotiate data packets comparing to the resources.

3.2 *VC creation at the RSU level*

The collection on RSU directories RD of all information containing ID resources with their attributes, and *Tl* determines the influence area of each supplier vehicle *Ai* [7] in which the supplier vehicle can offer its resources to consumers and transfers a registration packet to all RSUs that are in *Ai* to add registration packets to their RD enabling them to identify the supplier vehicle and their resources which belongs to Ai. If the supplier vehicle goes to another RSU and is nearest, Ai will change its value and send R2 (for example) another registration packet. Additionally, this latest development adds (updates) this packet to its RD and forwards it to all RSU within on Ai, for appending their RD.

The supplier vehicle which does not belong to a cluster periodically transmits a beacon to the closest RSU (every 2 seconds). These beacons empower RSUs to follow the supplier vehicle's travel. While the beacons are delayed arriving at RSU, they seek to identify the position of the supplier vehicle's area, which is called estimated area Ae. With this formula, we can measure the radius Re of $Ae = AV * (T c - T1)$, where T1 the time of sending the last beacon, Tc current time and AV is the average speed of the supplier vehicle. The RSU sends to all consumers the area Ae. It is meriting to note that the RSU will remove the ID of the supplier vehicle in its RD if does not receive beacons from it for a time $X = 10 * $ (interval beacon).

When a vehicle that does not belong to a cluster needs to consume some services. It formulates a REQUEST PACKET and sends it to its nearest RSU. To the desired resources and their attributes, the request packet contains the vehicle's geographic coordinates. If the RSU receives a request packet from the consumer vehicle, it searches its DR for one or more supplier vehicles that can face the vehicle's requirements.

The RSU that responds to a user's request will choose the corresponding supplier vehicle(s) from the candidate list Lc of supplier vehicles. Then, the RSU will formulate a RESPONSE PACKET, which contains the following elements for each supplier vehicle chosen by the RSU:

- The ID of the supplier vehicle.
- Latest supplier vehicle's locations from the latest beacon.
- Resources and their attributes.

The consumer vehicle defines the resources and their attributes that it requires from each supplier vehicle. Then it sends a SERVICE PACKET for each supplier vehicle picked using a routing protocol. The service packet includes the consumer vehicle's ID and their requests. Upon acquisition of the consumer vehicle's service packet, the supplier vehicle responds either with a NEGATIVE ACKNOWLEDGMENT if it cannot provide the user's request or its waiting line is busy, or with a SERVICE RESPONSE packet otherwise, by requesting the consumer vehicle for the payment method. The consumer and supplier vehicles can interchange data packets regarding the resources.

4 PERFORMANCE AND VALIDATION

4.1 *Simulation setup and parameters*

We used OMNeT++ 5.3 (2019) for the behavioral aspect and Sumo-0.32.0 (2019) as a mobility simulator to simulate FR-VC. FR-VC which is implemented in MANHATTEN grid 9×9, where we put 72 RSUs near to the junctions. The vehicle density (VD) is varied between 100 and 500 vehicles; each scenario of simulation is jointed with the value of VD (100, 300, 400, and 500). The details of the simulation parameters are shown in Table 1.

FR-VC is evaluated by taking these metrics:

1) Wait Time to Register (W_time): measures the time from sending the registration packet by supplier vehicle to be stored in VC (second).
2) Hit Percentage (HIT): percentage of requested services which is acknowledged by the RSU (%).

4.2 *Result and discussions*

To prove the performance of FR-VC, we performed an extensive comparison in terms of W_time and HIT comparing the same simulation parameters against three protocols, which are: CROWN (2013), DCCS-VC (2018) and TOPVISOR (2020).

As shown in Figure 1, FR-VC yielded better results than CROWN and TOPVISOR in terms of W_time. Compared to CROWN protocol, the outcome is justified by the use of the routing protocol

Table 1. Simulation parameters.

Parameter	Value
Simulation framework	Veins (OMNet++ and Sumo)
Mobility model	Manhattan
Simulation time	1000 s
Simulation area	9×9 km^2
Transmission range	500 m
Transfer rate	18 Mb/s
Vehicle density	[100–500] vehicles
Vehicles speed	Up to 70 km/h
Supplier vehicle density	1/4, 1/3 and 1/2 of vehicle density
The size of registration and request packet	128 Kbytes
Data Packet Size	[1–5] Kbytes
Supplier vehicle's queue	5 consumer vehicles

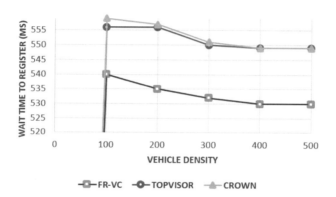

Figure 1. Performance evaluation comparisons of FR-VC with CROWN and TOPVISOR in terms of W_time.

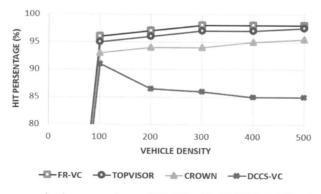

Figure 2. Performance evaluation comparisons of FR-VC with CROWN, DCCS-VC and TOPVISOR in terms of HIT.

E-GPSR in FR-VC that proves its efficiency. E-GPSR can deliver packets to reach RSUs better than CAN DELIVER (Mershad et al. 2012) which is used in CROWN. Regarding TOPVISOR, it's clear that it uses two levels to store packets at the VC, which are the RSUs and the central controllers, not like FR-VC that uses only RSUs to store packets.

As shown in Figure 2, CROWN and TOPVISOR protocols have ascending slopes except for the DCCS-VC protocol which has a descending slope. By varying the number of vehicles (100, 200,

300 and over 400), FR-VC performed better than CROWN, DCCS-VC and TOPVISOR in terms of HIT, the reason behind this best outcome can be substantiated by the non-correlation between the VD and this metric. It has a strong relationship with the critical location of RSUs in the map, where FR-VC chooses to deploy them near to junctions.

5 CONCLUSION

In this paper, we proposed a new approach to constructing vehicular clouds in urban areas, relying on mobile cluster-based vehicular clouds and RSU-based vehicular clouds, providing an opportunity to offer services to consumers by supplier vehicles for successful real-time consumption. The clustering technique that collaborates RSUs has noted a set of enhancements in performance assessment metrics terms. The simulation results proved the effectiveness of our proposal in terms of Wait Time to Register and Hit Percentage metrics. As future work, we will try to extend FR-VC to select the best services inside VC, hence investigating other metrics to give more credibility concerning comparison and evaluation.

REFERENCES

Brik, B., Lagraa, N., Tamani, N., Lakas, A. and Ghamri-Doudane, Y., 2018. Renting out cloud services in mobile vehicular cloud. *IEEE Transactions on Vehicular Technology*, *67*(10), pp. 9882–9895.

Dutta, R. and Thalore, R., 2017. A Review of Various Routing Protocols in VANET. *International Journal of Advanced Engineering Research and Science*, *4*(4), p. 237143.

Hasrouny, H., Samhat, A.E., Bassil, C. and Laouiti, A., 2017. VANet security challenges and solutions: A survey. *Vehicular Communications*, *7*, pp. 7–20.

Houmer, M. and Hasnaoui, M.L., 2019. An Enhancement of Greedy Perimeter Stateless Routing Protocol in VANET. *Procedia Computer Science*, *160*, pp. 101–108.

Jafari Kaleibar, F. and Abbaspour, M., 2020. TOPVISOR: Two-level controller-based approach for service advertisement and discovery in vehicular cloud network. *International Journal of Communication Systems*, *33*(3), p. e4197.

Li, L., Liu, J., Cheng, L., Qiu, S., Wang, W., Zhang, X. and Zhang, Z., 2018. Creditcoin: A privacy-preserving blockchain-based incentive announcement network for communications of smart vehicles. *IEEE Transactions on Intelligent Transportation Systems*, *19*(7), pp. 2204–2220.

Lo'ai, A.T., Bakheder, W. and Song, H., 2016, June. A mobile cloud computing model using the cloudlet scheme for big data applications. In *2016 IEEE First International Conference on Connected Health: Applications, Systems and Engineering Technologies (CHASE)* (pp. 73–77). IEEE.

Mershad, K. and Artail, H., 2013. Finding a STAR in a Vehicular Cloud. *IEEE Intelligent transportation systems magazine*, *5*(2), pp. 55–68.

Mershad, K., Artail, H. and Gerla, M., 2012. We can deliver messages to far vehicles. *IEEE Transactions on Intelligent Transportation Systems*, *13*(3), pp. 1099–1115.

Nagel, R. and Eichler, S., 2008, March. Efficient and realistic mobility and channel modeling for VANET scenarios using OMNeT++ and INET-framework. In *Proceedings of the 1st international conference on Simulation tools and techniques for communications, networks and systems & workshops* (pp. 1–8).

Omnet++ network simulation framework. "http://www.omnetpp.org", visited at 10/03/2019.

Qureshi, K.N., Bashir, F. and Iqbal, S., 2018, October. Cloud computing model for vehicular ad hoc networks. In *2018 IEEE 7th international Conference on Cloud Networking (CloudNet)* (pp. 1–3). IEEE.

Sumo, simulation for urban mobility. "http://sourceforge.net/apps/mediawiki/sumo", visited at 25/02/2019.

Innovative and Intelligent Technology-Based Services for Smart Environments – Ben Slama et al (eds)
© 2021 Taylor & Francis Group, London, ISBN 978-1-032-02030-3

Smart control algorithm in traffic lights at urban intersections

Chahrazad Hambli & Mourad Amad
LIMPAF Laboratory, Computer Sciences Department, Faculty of Sciences and Applied Sciences, Akli Mohand Oulhadj University, Bouira, Algeria

Nadjet Azzaoui
LINATI Laboratory, CSIT Department, Kasdi Merbah University, Ouargla, Algeria

ABSTRACT: Road congestion is a huge problem that people face most of the time, especially those who live in encumbered cities, which wastes time and money. Mainly, traditional traffic light control systems are far from providing effective urban traffic management in terms of journey time and waiting time and are unable to keep pace with the development of smart cities and urban mobility requirements. This article proposes an intelligent control algorithm of traffic lights. It is based on two important strategies, the first of which is the calculation of the shortest path to be performed between a source and a destination. The second strategy is an algorithm for managing and controlling traffic lights according to the number of vehicles on the road network studied. The number of vehicles is obtained using sensors placed on each road. Later, the number of vehicles is used to calculate the shortest path by applying an improved version of the Dijkstra algorithm. There is a controller who communicates with their neighbors in order to collect the necessary data and control the operation of traffic lights in an intersection. All the algorithms, including the calculation of the shortest path and the synchronization of the traffic lights are validated by traffic simulators that have been implemented using java language. The results show that the traffic light timing is calculated successfully.

Keywords: Internet of things, Sensors, Mobility, Intelligent Transportation Systems, Urban intersection, Road traffic control, Dijkstra algorithm.

1   INTRODUCTION AND MOTIVATIONS

While many aspects of our daily lives have become more enjoyable with the use of advanced technology, the transportation industry has taken a long time to catch up. Today, most cities face traffic jams and crippled traffic. This overload affects many areas of life, including the economy, industry, environment and health. This is due to the rapid growth of urban infrastructure and the increase in the number of people and transport vehicles. Traditional systems have found their limits, it has become necessary and imperative to develop a new intelligent traffic system that dynamically adapts to current traffic conditions. This new system faces the challenge of delivering more efficient and cost-effective results such as reducing congestion and maintaining the traffic system while protecting the environment by reducing toxic emissions (CO_2) and saving time.

Thus, to reduce urban traffic congestion, an intelligent dynamic control system for the synchronization of traffic lights is needed. And to provide a dynamic calculation, representative information in real time must be acquired. This information can be provided by sensor-sensor. In addition, there should be a controller as a computing unit that merges all the information from the sensor for the timing calculation. In addition, the more sensors there the more investment and maintenance is required. This paper considers using less sensors but with enough information to arrive to provide an effective decision.

                                                                           DOI 10.1201/9781003181545-23

Here, this paper proposes an intelligent control algorithm of traffic lights by calculating the shortest path on the road model. Sensors are required for each road. Vehicle-to-infrastructure and infrastructure-to-infrastructure communication is used to get the number of vehicles and the destination for later use to calculate the shortest path for vehicles. Each controller is connected with their neighbors and is used to control the traffic lights. In addition, the controller is used to collect data as a source of traffic information. Simulation experiments are conducted to examine the performance of our new algorithm using the java programming language and the ellipse development environment.

The rest of the article is organized as follows: section 2 provides an overview of the existing traffic light management models used. In Section 3, we describe how we proceeded using the Dijkstra algorithm to solve the traffic light management problem. In section 4, we discuss the results obtained. Finally, the conclusion and some perspectives for future work are given in section 5.

2 RELATED WORK

Among the studies that have been carried out on traffic light control, there are two types of approaches: That of traditional traffic controllers which consists either in pre-setting the synchronization of the green signal such that this cycle is repeated constantly regardless of the presence or absence of traffic demand (Mousavi 2017), or in the signal control actuated by the vehicle where the signal change depends on vehicles or other information on traffic demand (Mousavi 2017). The other part is adaptive signal control, which is also called real-time traffic control systems, in which the signal durations are automatically changed according to the current state of the intersection (i.e. traffic, queue length of vehicles in each lane of the intersection and fluctuation in traffic flow) (Mousavi 2017). This section is devoted to studies that evolve the effectiveness of current systems used in the management of traffic lights as follows:

Patel et al. (2016) proposes a model for minimizing and optimizing the total average control time and searches for the optimal green times in real time for heterogeneous road traffic without lanes. This model is based on an objective function which requires several parameters. In addition, it has been compared to the vehicle operated (VA) system which is part of the traditional traffic control systems. The performance measurements of this model are proven by simulation using the VISSIM simulator.

Hawi et al. (2017) presents a hybrid approach that uses fuzzy logic to aggregate the large amount of uncertain traffic data collected by a wireless sensor network (WSN) to determine the time of a green light based on the number of vehicles on the tracks. A number of vehicles corresponds to an interval defining a green light duration (ex. *less than 5 vehicles per minute gives the green light for 10 seconds*). This principle seems ideal to use: a simple theory applied to complex problems and the robustness of fuzzy control with respect to uncertainties. The drawbacks are nonetheless significant: the installation techniques and adjustments are empirical and there is no theory to demonstrate the stability and robustness of such a method.

Cao et al. (2017) proposes a traffic management framework based on multi-agent pheromones to reduce traffic congestion based on two traffic light control strategies, the dynamic vehicle rerouting strategy and the traffic light control strategy. The results of the simulation, which were analyzed, showed the improvement carried by this framework in terms of reduction of road congestion and pollution.

Javaid et al. (2018) proposes a combination of centralized and decentralized as being a hybrid approach to optimizing traffic on the roads, as well as an algorithm for the management of different traffic situations. This system manages traffic lights based on the density recovered through one of the cameras and sensors. Density is predicted using an algorithm based on artificial intelligence. In addition, RFIDs are used to apply priority management between emergency vehicles in the event of a traffic jam. The validity of this system has been proven by the development of a prototype.

Pratama et al. (2018) proposes adaptive traffic lights such that their synchronization checks are carried out based on the calculation of the density of road traffic. The synchronization control of the traffic lights is ensured by a set of algorithms and takes place within a server which is responsible for the collection of data and the control of the operations of the traffic lights, thus calculating the density using an image processing method. The validation of this study was carried out under real road conditions.

Jain et al. (2019) proposes a dynamic and automatic traffic signal management system approach, which is based on the arithmetic mean theorem. This approach uses six different parameters which are: the speed of the traffic units, the length of the queue, the time between arrivals between vehicles, the value of the centrality measurements and the expected value of the traffic congestion by the historical database where the output is the level of congestion. Subsequently, inference rules will be built based on the level of congestion and according to which the automatic update of the duration of the green and red lights will be carried out.

After studying many related works, it found that despite the availability of a huge proposed and ongoing contribution to solving traffic safety and efficiency issues with lower operational costs, the effective management of traffic lights has not been able to take place until now. This is due to the heaviness of the calculations, or the algorithms used require several parameters or many iterations, as well as the use of several hardware and software resources which increases the cost of solutions.

To contribute to the treatment of this problem caused by the bad management and the functioning of the traffic lights and to overcome the shortcomings of the associated work, we propose a new algorithm which is based on the use of the adapted Dijkstra algorithm and an effective strategy for the synchronization of traffic lights and thus the I2I and V2I communication as a link for the exchange of the necessary data in real time. The use of this algorithm is so light, it does not require a lot of resources and parameters, and the results are more efficient than those of the related works in measuring the optimization of vehicle travel times and to avoid saturation roads.

3   PROPOSED SOLUTION

We consider the road network model illustrated in the figure below (Figure 1). Each route is double-sided. As it contains intersections which are made up of different directions and each direction contains two lanes. The vehicle turning left uses the leftmost lane and the other lane is for vehicles going straight ahead or on the right. Each intersection has a controller that defines and applies the sequences of lights called cycle, traffic lights that control the traffic of vehicles, and a sensor in each direction and next to each light, which counts the number of vehicles that pass in each green light period (*when a vehicle enters a new direction the sensor adds the number of vehicles and when the latter leaves, it must be made to subtract*).

In our scenario, we must use two types of communication to ensure the exchange of necessary data and real-time traffic status information. Therefore, whenever a vehicle which is at point D wants to move to a specific point; it asks the point A controller to provide it with the shortest path to follow and this via a V2I type communication. Then, the point A controller (infrastructure) communicates with the other controllers (C, E and B) via an I2I type communication in order to retrieve information concerning the vehicles leaving in their direction.

The infrastructure is equipped with a wireless communication system by placing a dedicated controller at each intersection, the controller will then be able to better manage traffic lights. The controllers can therefore receive information from the sensors placed on the edges of the roadways. Thanks to this information, the controllers placed in the intersections must have the number of vehicles passed by each intersection to improve the flow of the path at their own intersection and must then communicate with each other (between intersections) in order to optimize the overall flow of the road traffic lane, and thus give the vehicle the best path calculated using the Dijkstra algorithm.

The Dijkstra algorithm is one of the best algorithms for calculating a shorter path in a graph where the weights are positive. Table 1 shows the modeling of our road network, which is represented in the form of a weighted graph:

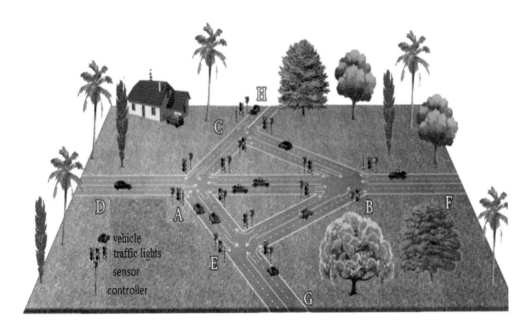

Figure 1. The road network studied.

Table 1. Road network modeling.

Vocabulary of the graphs	Road vocabulary
Graph (weighted)	Road network.
Summit	City
Arc	One-way road
Weight	Congestion_rate=Number_vehiculs/Road_capacity

In our case, the principle of this algorithm is to calculate the shortest path by modifying the weight of the edges of the original algorithm by:

$$Congestion_{rate} = \frac{Number\_vehicules}{Road\_capacity} \tag{1}$$

When:

Number_vehicules is number of vehicles in a section.
Road_capacity is the number of vehicles that can be supported in a section.

The management and control algorithm of traffic lights at the intersection level does not depend on the duration of the lights but on the number of vehicles taking the road in each direction such that each controller must check the state of the direction next to allow the vehicle to pass.

Our proposal is to reduce the congestion problem by optimizing the traffic lights based on the Dijkstra algorithm to determine the shortest path in a given instant and an algorithm for the management of intelligent traffic lights (dynamic traffic lights) where the management of the intersection does not depend on the time of the green lights but on the number of vehicles using the road in each direction. The working diagram of our algorithm is shown in Figure 2.

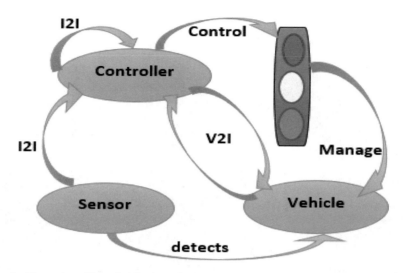

Figure 2. An illustration of how the algorithm works and the interactions between components of the road network.

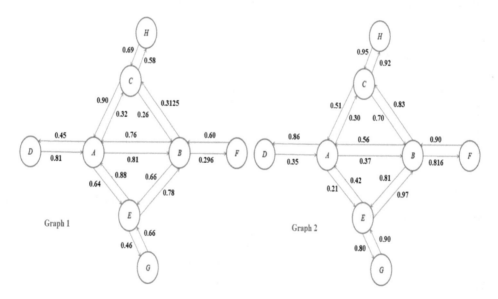

Figure 3. The congestion rate of the road network.

4 SIMULATION AND RESULT ANALYSIS

The algorithm for intelligent control of traffic lights at urban intersections was tested by simulation with the java programming language. This simulation is based on the road network shown in Figure 1. The results obtained from this experiment are the congestion rates of each direction (represented by a weighted graph) and the shortest path from point D to point F. The calculation of the traffic congestion rate on all roads is applied cyclically by the controllers (installed at intersections). The results of the experiment can be seen in Figures 3–5.

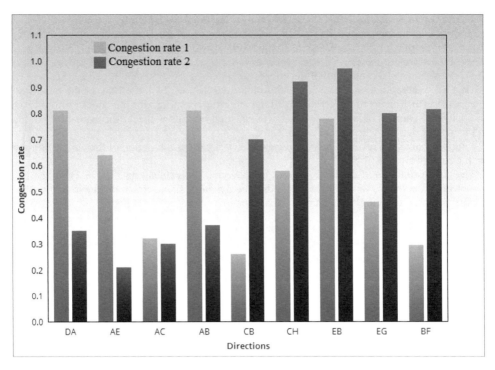

Figure 4.   The congestion rate of the road network in the directions from D to F.

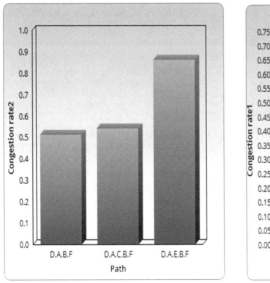
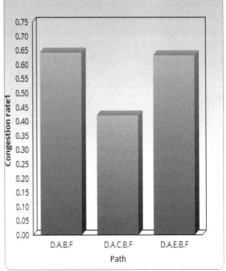

Figure 5.   Best path from D to F according to congestion rate.

Figure 3 represents the stage of construction of the graph weighted by the congestion rates of each direction, and thereafter this graph will be the input of the Dijkstra algorithm. This figure shows two graphs: graph 1 is the state of the road network at time T1 and graph 2 is the state of the road network at time T2.

Figure 4 shows the evaluation of the congestion rate of each direction in two consecutive times T1 and T2, where there are directions with high rates at time T1 but on the contrary at time T2 have a low rate. Thanks to the efficiency of the algorithm proposed for the control of lights which calculates the shortest path according to the congestion rate and it checks the availability of places in the next direction so that vehicles can enter.

Figure 5 shows the result of calculating the shortest path as a function of the congestion rate in each direction at time T1 and T2.

The congestion rate of the road network is between 0 (no vehicle) and 100%. From the results of several simulations, we can summarize that the dynamic management of traffic lights at urban intersections by applying the version of the improved algorithm of Dijkstra and the traffic light control algorithm that has brought improvements in terms of congestion reduction and satisfactory intersection control. The improved Dijkstra algorithm on the road network weighted by the congestion rate of each direction instead of the distance and the traffic light control strategy depends on the number of vehicles in the destination direction. Then, the calculation of the shortest path was successfully applied to ensure the minimization of the lost time. In Figure 5, we can see that the shortest path is proportional to the congestion rate of the road network.

5   CONCLUSION

The appropriate solution is to find control algorithms to avoid the appearance of traffic congestion. In this article, we presented a solution for managing and controlling traffic signals in a smart city with the aim of streamlining road traffic. This is done through an approach based on a V2I and I2I communication and an algorithm to find the best way to reduce travel time. The results obtained show that the intelligent control algorithm of traffic lights at urban intersections proposed in this study was available for effective control of traffic lights. furthermore, the congestion rate value has been used successfully to calculate the shortest path from a source to a destination. Additionally, communication and the exchange of data and information between controllers has been used to improve performance. For future work, we aim to enrich our approach by adapting traffic lights to the presence of priority vehicles (ex. ambulance, etc.).

REFERENCES

Cao, Z., Jiang, S., Zhang, J., and Guo, H. (2017, JULY). A Unified Framework for Vehicle Rerouting and Traffic Light Control to Reduce Traffic Congestion. IEEE TRANSACTIONS ON INTELLIGENT TRANSPORTATION SYSTEMS, VOL.18, NO.7.

Hawi, R., Okeyo, G., and Kimwele, M. (2017). Smart Traffic Light Control using Fuzzy Logic and Wireless Sensor Network. in *2017 Computing Conference*, London, United Kingdom.

Jain, A., Yadav, S., Vij, S., Kumar, Y., and Tayal, D.K. (2019, September). A Novel Self-Organizing Approach to Automatic Traffic Light Management System for Road Traffic Network. Wireless Pers Commun 110, (pp. 1303–1321). https://doi.org/10.1007/s11277-019-06787-z.

Javaid, S., Sufian, A., Pervaiz, S., and Tanveer, M. (2018, February). Technology (ICACT) Smart Traffic Management System Using Internet of Things. International Conference on Advanced Communications.

Mousavi, S. S., Schukat, M., & Howley, E. (2017). Traffic light control using deep policy-gradient and value-function-based reinforcement learning. IET Intelligent Transport Systems, 11(7), (pp. 417–423).

Patel, A., Mathew, T. V., & Venkateswaran, J. (2016). Real-time adaptive signal controller for non-lane following heterogeneous road traffic. 2016 8th International Conference on Communication Systems and Networks (COMSNETS).

Pratama, B., Christanto, J., Hadyantama, M.T., and Mui, A. (2018, October). Adaptive Traffic Lights through Traffic Density Calculation on Road Pattern. Conference Paper in iCAST1.

Innovative and Intelligent Technology-Based Services for Smart
Environments – Ben Slama et al (eds)
© 2021 Taylor & Francis Group, London, ISBN 978-1-032-02030-3

Speed control for BLDC motor using fuzzy sliding mode

K. Cherif
*Laboratory of Automatic (LARA) National Engineering School of Tunis, Tunis El Manar University, Tunis,
Tunisia*

A. Sahbani
*Laboratory of Automatic (LARA) National Engineering School of Tunis, Tunis El Manar University, Tunis,
Tunisia*
Medina College of Technology Al Madinah Al Munawwarah, Kingdom of Saudi Arabia

K. Ben Saad
*Laboratory of Automatic (LARA) National Engineering School of Tunis, Tunis El Manar University, Tunis,
Tunisia*

ABSTRACT: This paper presents the elaboration of a fuzzy sliding mode controller (FSMC) used
for a Brushless DC motor (BLDC). The studied controller specifies modifications in the output
control signal by using the sliding surface and its variation in order to guarantee the attraction
conditions and the sliding mode stability. The proposed FSMC is implemented in Matlab Simulink
and compared to a conventional PI control. The simulation results confirm the efficiency and the
robustness of the proposed FSMC strategy against motor parameters and load variations.

1   INTRODUCTION

Brushless DC motors (BLDC) are increasingly used in several application areas, such as motion
control systems, variable speed drives, aerospace actuators, computer peripherals (DVD players),
compressors, automotive applications, digital cameras, medical scanners, small fans and motor
pumps and HVAC (Heat, Ventilation and Air-Conditioning) applications. Because of the high
starting torque, high speed and high efficiency compared with other motors and especially the
reliability and low maintenance of BLDC motors, they are preferred for electric vehicle applications
[1–4]. Since the first models of electric vehicles, the basic elements making up the drive train have
been based on accumulators to store electrical energy, a motor to transform electrical energy into
mechanical energy and an electrical converter to transfer and control the power from the batteries to
the motor. These converters provide the power distribution function during energy exchanges [2].
BLDC motors are one type of permanent magnet motors. Sinusoidal flux distribution (sinusoidal
e.m.f.) or trapezoidal flux distribution (trapezoidal e.m.f.) are possible with BLDC motors. BLDC
motors are an attractive candidate for sensorless electric drives with trapezoidal FEM since only
two of the motor windings are energized at each cycle period.

Starting these motors, involves knowledge of the exact location of the rotor (as in the case
of synchronous motors) in order to correctly supply the corresponding phases and to achieve
the maximum torque. Within this context, it is envisaged to develop appropriate strategies for
these systems, using intelligent control techniques to improve performance. There is a diversity of
published papers in relation to the control of the BLDC motors speed used for EV. They realize the
motor speed control using a conventional PI controller [5–8], authors show that motor performance
can lead to torque disturbances due to parameter variations and sensitivity to the uncertain nature
of the system. Other authors adjust the PID parameters using genetic algorithm, fuzzy algorithm
[9–10], or using a fractional order PID controller. In [12] the authors show that the BLDC motor
is usually nonlinear and there are a variety of disruptions, to solve these limitations and to improve
they propose a Sliding Mode Control (SMC). Although SMC is vigorous in the motorchanges,

DOI 10.1201/9781003181545-24   163 the increase in gain used to monitor the target causes an unnecessary chattering effect and causes ripples in the responses, as well as high frequency switching in the converters [11]. To solve and remove the chattering phenomena, authors [11] propose a fuzzy sliding mode control applied to control the output voltage in DC- DC converter.

In this paper we applied this idea to control the speed of BLDC motor used for Ev application.

Section 2 of this work presents a general description, the operation principle and the modeling of the BLDC motor, the proposed fuzzy sliding mode control is presented in section 3, section 4 present the simulation results, and finally the conclusion is presented in section 5.

2 BLDC MOTOR

2.1 *General description*

The BLDC motor is a synchronous motor, meaning that the stator-created magnetic field and the rotor-created magnetic field rotate at the same frequency[18].The brushless motor is, as its name suggests, a brushless motor similar to a DC motor whose switching is performed electronically (electronic converter) rather than mechanically (brushes and rings). This removes the mechanical commuting problems: sparks and wear on the assembly of the commutator-brush, making the motor more stable compared to a DC motor.

In this case, the stator consists of excitation coils that are normally a multiple of 3 and are connected most frequently in a star configuration and may also be connected in a delta configuration. Most BLDC engines often have a series of three Hall-effect sensors that make it possible to know the direction of the rotor, located at 60° or 120° from each other. Knowing the location of the rotor helps the electronic circuit to turn on the power supply.

2.2 *Operating principle and mathematical modeling of BLDC motor*

The operating principle remains unchanged since the coils are connected in series in a BLDC motor, so that there are always three phases and a common neutral.

The BLDC motor operates from three variable voltage sources supplied by an inverter, creating a rotating magnetic field in the stator. Equipped with a permanent magnet, the rotor appears to obey the magnetic field that rotates. PWM methods are added to control the voltage magnitude in order to control the speed.

Consider a brushless motor with star-connected stator windings, and the three stages are symmetrical and similar.

There is no change in the reticence of the engine due to the non-bleeding of the rotor.

The electrical equations that are linked to the stator in a fixed reference frame are defined by:

$$\begin{bmatrix} V_{as} \\ V_{bs} \\ V_{cs} \end{bmatrix} = \begin{bmatrix} R_s & 0 & 0 \\ 0 & R_s & 0 \\ 0 & 0 & R_s \end{bmatrix} \begin{bmatrix} i_a \\ i_b \\ i_c \end{bmatrix} + \frac{d}{dt} \begin{bmatrix} L_{aa} & L_{ab} & L_{ac} \\ L_{ba} & L_{bb} & L_{bc} \\ L_{ca} & L_{cb} & L_{cc} \end{bmatrix} \begin{bmatrix} i_a \\ i_b \\ i_c \end{bmatrix} + \begin{bmatrix} e_a \\ e_b \\ e_c \end{bmatrix} \qquad (1)$$

where:
V_{as}, V_{bs} et V_{cs}: Stator voltages.
  R_s: The stator resistance of a phase
  i_a, i_b et i_c: The stator currents.
  L_{aa}, L_{bb} et L_{cc}: The self-inductances of phases a, b and c.
  L_{ab}, L_{ac} et L_{cb}: The mutual inductances between the phases.
  e_a, e_b et e_c: The electromotive forces of the phases.
  The inductances of the different motor windings and their reciprocal inductances are equivalent, hence the following relationships:

$$L_{aa} = L_{bb} = L_{cc} = L \qquad (2)$$

$$L_{ab} = L_{ba} = L_{ac} = L_{ca} = L_{bc} = L_{cb} = M \tag{3}$$

Replacing equations (2) and (3) in (1), we obtain the following model:

$$\begin{bmatrix} V_{as} \\ V_{bs} \\ V_{cs} \end{bmatrix} = \begin{bmatrix} R_s & 0 & 0 \\ 0 & R_s & 0 \\ 0 & 0 & R_s \end{bmatrix} \begin{bmatrix} i_a \\ i_b \\ i_c \end{bmatrix} + \frac{d}{dt} \begin{bmatrix} L & M & M \\ M & L & M \\ M & M & L \end{bmatrix} \begin{bmatrix} i_a \\ i_b \\ i_c \end{bmatrix} + \begin{bmatrix} e_a \\ e_b \\ e_c \end{bmatrix} \tag{4}$$

The stator currents are balanced as follows $i_a + i_b + i_c = 0$
As follows, this simplifies the inductance matrix

$$Mi_b + Mi_c = -Mi_a \tag{5}$$

In this way, the mathematical model becomes:

$$\begin{bmatrix} V_{as} \\ V_{bs} \\ V_{cs} \end{bmatrix} = \begin{bmatrix} R_s & 0 & 0 \\ 0 & R_s & 0 \\ 0 & 0 & R_s \end{bmatrix} \begin{bmatrix} i_a \\ i_b \\ i_c \end{bmatrix} + \frac{d}{dt} \begin{bmatrix} L-M & 0 & 0 \\ 0 & L-M & 0 \\ 0 & 0 & L-M \end{bmatrix} \begin{bmatrix} i_a \\ i_b \\ i_c \end{bmatrix} + \begin{bmatrix} e_a \\ e_b \\ e_c \end{bmatrix} \tag{6}$$

The trapezoidal- shaped electromotive powers. The following equation describes them:

$$\begin{bmatrix} e_a \\ e_b \\ e_c \end{bmatrix} = \omega_m \lambda_m \begin{bmatrix} f_{as}(\theta_r) \\ f_{bs}(\theta_r) \\ f_{cs}(\theta_r) \end{bmatrix} \tag{7}$$

where:
ω_m: Angular velocity in rad/s
θ_r: The rotor position in rad
$f_{as}(\theta_r), f_{bs}(\theta_r)$ and $f_{cs}(\theta_r)$ are functions with the same trapezoidal shape as electromotive forces.
The electromechanical torque is defined by:

$$C_e = [e_a i_a + e_a i_a + e_a i_a] / \omega_m \tag{8}$$

The mechanical equation is therefore:

$$J \frac{d\omega_m}{dt} + B\omega_m = C_e - C_r \tag{9}$$

where:
J: Moment of inertia of the rotating part (kg·m²)
B: Viscous friction coefficient (N·m·s/rad)
C_r: Resistant torque (N·m)
As previously described, an electronic switch is required for the BLDC motor to ensure successive power supply to the windings. Figure 1 shows the schematic diagram of this electronic switches (K_1, K_2, K_3, K_4, K_5 and K_6) connected to the stator of the BLDC motor.

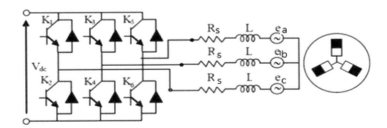

Figure 1.   Electronic switch connected to BLDC motor.

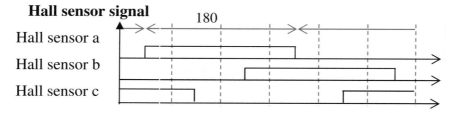

Figure 2.   Hall sensor signal function of rotor position (degree).

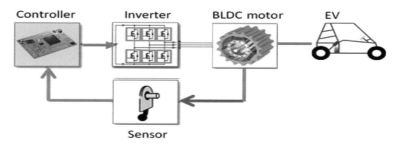

Figure 3.   The control block of BLDC motor.

The three BLDC motor phases are entirely symmetrical and generate periodically induced trape-zoidal shaped FEMs. The rotor location signals provided by the Hall Effect Sensor, as shown in Figure 2, provide the control sequences of the individual electronic switch switches.

3   SPEED CONTROL OF BLDC MOTOR

3.1   *System description*

Figure 3 shows the control block of the BLDC motor associated with an inverter used as an electric drive in EV. The speed sensor and the rotor position are taken from the three hall sensors. The feedback information is used to control the speed of this motor.

The controller unit obtains the desired speed and the real speed of the motor, and also the three hall sensor signals Hsa Hsb and Hsc, then, generates a PWM in order to control the output voltage of the inverter and then the speed of the motor [].

3.2   *PI control*

To control the speed the BLDC motor, we will include a parallel proportional-integral form corrector in this portion described by:

$$u(t) = k_p e(t) + \frac{1}{T_i} \int_{-\infty}^{t} e(t)d\tau \tag{10}$$

Whose function for transfer is defined as follows:

$$\frac{U(p)}{E(p)} = k_p + \frac{1}{T_i p} \tag{11}$$

Figures 4 and 5 present respectively, the simulation results of the speed and current response using PI controller.

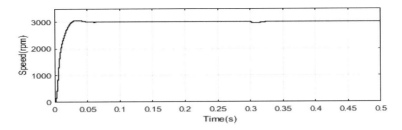

Figure 4.   Speed responses using PI controller.

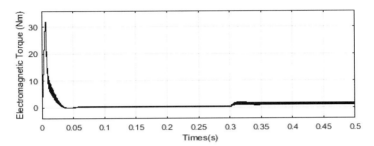

Figure 5.   Torque responses using PI controller.

3.3   *Fuzzy Sliding Mode Control (FSMC)*

The first stage to build a FSMC is to calculate the sliding surface, let us propose the sliding surface S as follow [12]:

$$S(x) = e(x) = \omega_m^* - \omega_m \tag{12}$$

where e is the BLDC speed error

In sliding mode control, the control signal u is the combination of a nonlinear control U_n and an equivalent control U_{eq}.

the control signal u is [12]:

$$u = C_e = U_{eq} + U_n \tag{13}$$

Let

$$\dot{S}(x) = \dot{e}(x) = \frac{d\omega_m^*}{dt} - \frac{d\omega_m}{dt} = 0 \tag{14}$$

From the model of the developed BLDC engine, we get:

$$\frac{d\omega_m}{dt} = \frac{C_e - C_r - B\omega_m}{J} \tag{15}$$

Then,

$$\dot{S}(x) = \frac{d\omega_m^*}{dt} - \frac{d\omega_m}{dt} = \dot{\omega}_m^* - \frac{U_{eq} - C_r - B\omega_m}{J} \tag{16}$$

The equivalent control signal is:

$$U_{eq} = J\dot{\omega}_m^* + C_r + B\omega_m \tag{17}$$

167

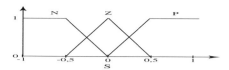

Figure 6.   Surface S membership functions.

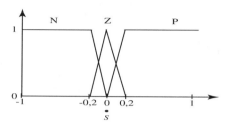

Figure 7.   Surface change $\dot{S}$ membership functions.

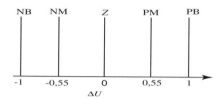

Figure 8.   Output singletons.

Table 1.   The rule base.

	S	N	Z	P
$\dot{S}$	P	Z	PM	PB
	Z	NM	Z	PM
	N	NB	NM	Z

The discontinuous components of the control signal is:

$$U_n = \varepsilon\, sign(s) + ks \tag{18}$$

Therefore, the control signa u is:

$$U = U_{eq} + U_n = J\dot{\omega}_m^* + C_r + B\omega_m + \varepsilon sign\,(s) + ks \tag{19}$$

The stability condition is satisfied if

$$\dot{S}(x)\,S(x) = (-\varepsilon\, sign(s) - ks)\,s \leq 0 \tag{20}$$

The proposed FSMC uses the surface S and its variation $\dot{S}$ as inputs to determine the changes on the output control signal $\Delta U(K)$ to force the stability condition to be established. Thus, the rule are defined in order to satisfy the inequality (20).

Trapezoidal and triangular membership functions, labelled by P (Positive), Z (Zero), and N (Negative) were chosen for the surface and its variation. Figure 6 and Figure 7 represent these functions respectively. Figure 8 represents the fives normalized singletons used for the output

168

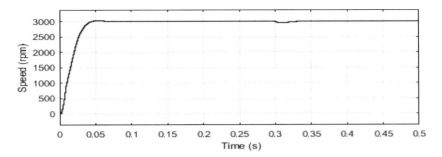

Figure 9.   Speed responses using FSMC.

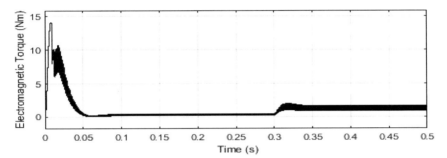

Figure 10.   Torque responses using FSMC.

signal, which are denoted PB (Positive Big), PM (Positive Middle), Z (Zero), NM (Negative Middle) and NB (Negative Big).

Table 1 gives the rule base for the proposed FSMC.

4   SIMULATION RESULTS

The studied FSMC was verified by simulation. Figure 9 presents the simulated voltage speed response of the BLDC motor for 3000-rpm reference speed. Figure 10 shows the electromagnetic torque. From the simulation results, we can conclude that FSMC is robust and removes quickly the perturbation caused by an increase in load torque at t=0.3 s after a short time.

5   CONCLUSION

This paper presents two control techniques for a BLDC motor. The first technique is a PI controller. The second control method is a FSMC. The combination of sliding mode control and fuzzy control aims to improve the transient response of the speed by decreasing the speed and motor current overshoot. The simulation results show that the application of FSMC to control the speed of BLDC motor is more robust in the case of the load and speed.

REFERENCES

[1] MA. Hassanin, F. Abdel-Kader, S. Amer et al., "Operation of Brushless DC Motor to Drive the Electric Vehicle." In 2018 Twentieth International Middle East Power Systems Conference (MEPCON) (pp. 500–503). IEEE, 2018.

[2] X. Nian, F, Peng,, & H. Zhang, "Regenerative braking system of electric vehicle driven by brushless DC motor," IEEE Transactions on Industrial Electronics, 61(10), 5798–5808, 2014.

[3] C. Chhlonh, DC. Riawan, & H.Suryoatmojo, "Modeling and Simulation of Independent Speed Steering Control for Front In-wheel in EV Using BLDC Motor in MATLAB GUI." In 2019 International Seminar on Intelligent Technology and Its Applications (ISITIA) (pp. 270–275). IEEE, 2019.

[4] HF. Prasetyo, AS. Rohman, FI. Hariadi, & H. Hindersah, "Controls of BLDC motors in electric vehicle Testing Simulator." In 2016 6th International Conference on System Engineering and Technology (ICSET) (pp. 173–178). IEEE, 2016.

[5] HS. Hameed, "Brushless DC motor controller design using MATLAB applications." In 2018 1st International Scientific Conference of Engineering Sciences-3rd Scientific Conference of Engineering Science (ISCES) (pp. 44–49). IEEE, 2018.

[6] LK. Agrawal, BK. Chauhan, & GK. Banerjee, "Speed Control of Brushless DC Motor Using Conventional Controllers. International Journal of Pure and Applied Mathematics," 119(16), 3955–3961, 2018.

[7] A. Rawat, M. Bilal, & MF. Azeem, "PI Controller Based Performance Analysis of Brushless DC Motor, Utilizing MATLAB Simulink Environment. Journal of Scientific Research and Reports," 38–43, 2020.

[8] R. Rakhmawati, FD. Murdianto, & MW. Alim, "Soft Starting & Performance Evaluation of PI Speed Controller for Brushless DC Motor Using Three Phase Six Step Inverter." In 2018 International Seminar on Application for Technology of Information and Communication (pp. 121–126). IEEE, 2018.

[9] AH. Ahmed, B.Abd El Samie, & AM. Ali, "Comparison between fuzzy logic and PI control for the speed of BLDC motor." International Journal of Power Electronics and Drive Systems, 9(3), 1116, 2018.

[10] LK. Agrawal, BK. Chauhan, & GK. Banerjee, "Speed control of brushless DC motor using Fuzzy controller." International Journal of Pure and Applied Mathematics, 119(15), 2689–2696, 2018.

[11] A.Sahbani, K.Ben Saad and M.Benrejeb, Chattering phenomenon suppression of Buck-Boost DC-DC converter with Fuzzy Sliding Modes control, International Journal of Electrical and Electronics Engineering (IJEEE), 4:1, pp.1–6, 2010.

[12] Y. Shao, R. Yang, J. Guo, and Y. Fu, "Sliding mode speed control for brushless DC motor based on sliding mode torque observer," in Proc. IEEE Int. Conf. Inf. Autom., Lijiang, China, 2015, pp. 2466–2470.

Section III: Green energy production and transfer systems

Innovative and Intelligent Technology-Based Services for Smart
Environments – Ben Slama et al (eds)
© 2021 Taylor & Francis Group, London, ISBN 978-1-032-02030-3

Load profile management effect on a stand-alone photovoltaic system multi-criterion design optimization

S. Fezai & J. Belhadj
ENSIT, Université de Tunis, Montfleury, Tunisia
Laboratoire des Systèmes Electriques, Ecole Nationale d'Ingénieurs de Tunis,
Université de Tunis El Manar, Tunis le Belvédère, Tunis, Tunisia

ABSTRACT: In this paper, a load profile management procedure is proposed and evaluated by the bi-objective design optimization of a Stand-Alone Photovoltaic (SAPV) system with battery storage. Load shifting and amplitude modulation are applied on the load profile shape for different rates of Loss of Load Probability (LLP) referring to consumers' preferences. The Life Cycle Cost (LCC) and the Loss of Power Supply Probability (LPSP) are the optimization objectives presenting the economic criterion and system reliability evaluation criterion. The bi-objective optimization using the genetic algorithm (GA) aims computing the optimal combination of PV panels and batteries subject to the constraints related to the battery bank State of Charge (SOC). The energetic modeling of the system's components enabling energy flows simulation is presented. Results showed the significant effect of the load profile management on the system design. Furthermore, the optimal solution choice depends on consumers' preferences regarding the system's economic and reliability aspects.

1 INTRODUCTION

Renewable energy is the fastest growing source of energy and is becoming the largest source of power by 2040. This rapid growth is aided by falling costs of clean energy technologies. The share of renewable energies in total power generation will reach 40%. Wind and solar photovoltaics (PV) bring a major source of affordable, low-emissions electricity which presents an impressive solution to the dual challenge related to carbon emissions reduction and the rising energy demand. The electrical energy consumption growth is participating in the rise of renewable energy systems implementation, especially photovoltaic (PV) production. Solar PV have become the most competitive option for electricity for residential and commercial applications. It plays a significant and growing role in electricity generation (Ren21 2019; IEA 2018; BP 2019; Solar Power Europe 2017).

The management between the intermittent generation periods of PV resource and consumption periods is very convoluted in SAPV systems. Therefore, the load profile major impact on the system's design and reliability was investigated for different optimization criteria. Mainly, economic cost minimization and higher system reliability are considered optimization objectives. Generally considered as an optimization objective, the SAPV system reliability evaluation criterion is the LPSP. Researchers also combined multiple optimization objectives in order to find a compromise solution as in Zhou et al. (2010), Jain & Jain (2017). The most common sizing optimization criterion is the economic aspect as presented in Kartite & Cherkaoui (2016), Okoye & Solyali 2017), Abdulaziz et al. (2017). It consists in computing the least expensive system configuration which achieves the required autonomy level. In (Thiaux et al. 2010), the load profile impact of SAPV systems was studied with the gross energy requirement as the criterion. A study of the influence of load profile variation on the optimal sizing of a stand-alone hybrid

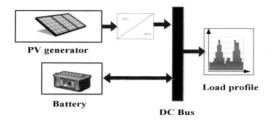

Figure 1.   SAPV system.

Table 1.   Characteristics of the PV panel and the battery.

PV panel	Maximum power (W)	95
	Peak-point voltage of maximum power (V)	18.43
	Peak-point current of maximum power (A)	5.16
	Open circuit voltage (V)	21.98
	Short-circuit current (A)	5.64
	Surface area (m^2)	0.66
Battery	Nominal voltage (V)	12
	Nominal capacity (Ah)	100
	Round trip efficiency (%)	85
	Depth of discharge (DOD) (%)	80

PV/Wind/Battery/Diesel system was investigated in Ould Bilal et al. (2013). Optimal sizing design and energy management of stand-alone photovoltaic/wind generator systems was presented in Tégani et al. (2014).

In this context, the load profile management effect on the bi-objective optimal design of a SAPV system using GA is presented. The LCC and LPSP are the optimization criteria. The load management procedure encloses time shifting, amplitude modulation and load shedding. The LLP is a criterion referring to the modification rate applied on the load profile. The remainder of this paper is organized as follows: in section 2, the SAPV system architecture and modeling is presented. Section 3 is dedicated to the optimization problem formulation. The load profile management procedure is detailed in section 4. Sections 5 and 6 are respectively dedicated to results analysis and conclusion.

2   SAPV SYSTEM ARCHITECTURE AND MODELING

2.1   *SAPV system architecture*

The studied SAPV system's architecture is presented in Figure 1. A PV generator is considered to be the energy production resource and the battery bank ensures surplus energy storage and load supply in case of PV panels' lack of production. The characteristics of the PV panel and the battery are presented in Table1.

This system is dedicated to supplying the daily load profile presented in Figure 2. The complexity of the studied load profile resides in the consumption peaks related to the inhabitant's attitudes. The first consumption peak supply is mostly ensured by the battery bank. Then, the PV generator ensures both consumer supply and battery banks are charging. The second consumption peak is also provided by the battery bank as the PV generation notably decreases.

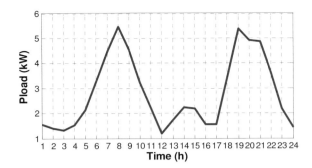

Figure 2.   Daily load demand profile.

2.2   *SAPV system modeling*

For a further system's optimization energetic modeling was considered. It permits the simulation of energy flows along the simulation time (Thiaux et al. 2010).

2.2.1   *PV production modeling*

The output power of the PV system is obtained from the solar radiation data I_r, the PV area A_{PV} and the power conversion efficiency of PV panels η_{PV}. It is expressed as follows (Cherif & Belhadj 2016):

$$P_{PV} = \eta_{PV} \times A_{PV} \times I_r \tag{1}$$

$$\eta_{PV} = \eta_{mp,STC} \times \left[1 - \beta \times \left(T_C - T_{C,NOCT} \right) \right] \tag{2}$$

where $\eta_{mp,STC}$ is the maximum power point efficiency under Standard Test Conditions (STC: 1 kW/m^2, 25°C $\pm$ 2°C); β is the generator efficiency temperature coefficient (0.005 per °C); $T_{C,NOCT}$ is the Normal Operating Cell Temperature when PV cell operates under STC ($T_{C,NOCT}$: 47 $\pm$ 2°C); and T_C is the cell temperature which is expressed by the following empirical equation:

$$T_C = 30 + 0.0175 \times (I_r - 300) + 1.14 \times (T_a - 25) \tag{3}$$

where T_a is the ambient temperature.

2.2.2   *Battery storage modeling*

An ideal lead-acid battery is used due to its technical maturity and its economic cost. The charging and discharging process are expressed respectively by equations 4 and 5 (Deshmukh & Deshmukh 2008; Wu et al. 2011; Cherif & Belhadj 2016; Fezai & Belhadj 2016).

$$E_B(t) = E_B(t-1) + (E_G(t) - E_{CONS}(t)) \times \eta_{Bat} \tag{4}$$

$$E_B(t) = E_B(t-1) - (E_{CONS}(t) - E_G(t)) \tag{5}$$

Where E_B is the batteries stored energy; E_G is the generated PV energy; E_{CONS} is the energy load demand; and η_{Bat} is the battery round trip efficiency.

Battery charging and discharging limitations involve two constraints related to the SOC as expressed by:

$$SOC_{min} \leq SOC(t) \leq SOC_{max} \tag{6}$$

3 SIZING OPTIMIZATION OF THE SAPV SYSTEM

In order to evaluate the effect of the load profile changes on the SAPV system's behavior, two main features will be focused in this paper: the LCC as the economic criterion and the energy dissatisfaction ratio LPSP. The decision variables are the number of PV panels (Npv) and the number of batteries (Nbat). The GA, which is a stochastic optimization technique, is used for the bi-objective sizing optimization of the SAPV system. The MATLAB multi-objective optimization toolbox "optimtool" was used. The optimization problem formulation is expressed as follows:
  Minimize:

$$LCC(N_{PV}, N_{Bat}) = LCC_{PV} + LCC_{Bat} \tag{7}$$

$$LPSP(N_{PV}, N_{Bat}) = \frac{\sum_{t=1}^{t\,max} LPSt(t)}{\sum_{t=1}^{t\,max} E_{Cons}(t)} \tag{8}$$

Subject to:

$$g_1 = 20\% - \min(SOC(t)) \leq 0 \tag{9}$$

$$g_2 = \max(SOC(t)) - 100\% \leq 0 \tag{10}$$

$$N_{PV\,min} \leq N_{PV} \leq N_{PV\,max} \tag{11}$$

$$N_{Bat\,min} \leq N_{Bat} \leq N_{Bat\,max} \tag{12}$$

where LCC_{PV} is the economic model of PV system (poly-Si) expressed as a function of PV area (A_{PV}) (Cherif & Belhadj 2016):

$$LCC_{PV} = 440.13 \times A_{PV} + 96.1 \tag{13}$$

LCC_{Bat} is the economic model of batteries expressed as a function of nominal capacity (C_{Bat}) (Cherif & Belhadj 2016):

$$LCC_{Bat} = 11 \times C_{Bat} + 146.8 \tag{14}$$

LPSt (t) is the energy deficit calculated during system simulation when the battery bank SOC reaches its minimum allowable value SOCmin. Equation (9) and (10) present the constraints related to the battery bank SOC. The lower and higher acceptable SOC rates are respectively 20% and 100%. Equation (11) and (12) are the decision variable bounds constraints.

4 LOAD PROFILE MANAGEMENT PROCEDURE

A typical load profile is usually composed of different load types: permanent, primary (or essential) and secondary loads. Load management consists in increasing the correlation between the PV production and the load profile by acting on the load profile shape. The load management procedure, for a fixed number of PV panel and batteries, is detailed in the flowchart in Figure 3. Two different actions can be applied on the load profile shapes: amplitude modulation and load shifting. An amplitude modulation is not only a peak clipping, it also consists of replacing the same energy amount during a specific time period, T, by a longer time, Tref (Tref >T). This type of change can be applied on secondary loads in case of supply shortages. The objective of load time shifting is to bring closer the first consumption peak to PV production peak in an attempt to attain the ideal "solar" consumer (Thiaux et al. 2010).

The management process takes into account available energy and battery bank SOC. The produced PV energy can be used to supply consumers during the first consumption peak using load

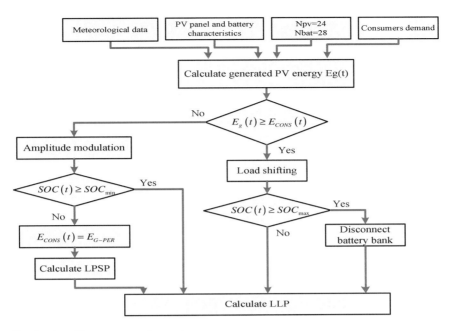

Figure 3. Load profile management procedure.

shifting procedure. The battery bank is disconnected in case the SOC reaches the upper bound SOCmax. Load modulation can be devoted to the second consumption peak due to the low provided energy level. Therefore, only permanent loads would be supplied. LLP is a factor related to the load profile management procedure and consumers preferences. Energy shed is calculated whether it's amplitude modulation or time shifting. LLP is expressed by the following equation (Pillai 2008; Bataineh 2012; Pilai 2014).

$$LLP(\%) = \frac{\sum_{t=1}^{t\max} \left(E_{cons}(t) - E_{c\_man}(t)\right)}{\sum_{t=1}^{t\max} E_{cons}(t)} \tag{15}$$

Where E_{c_man} is the modified load profile.

5 RESULTS ANALYSIS

The bi-objective optimization results for various values of LLP are illustrated in Figure 4. It presents a set of solutions for different values of LPSP and LCC. The optimal solution varies with LLP rate.

The optimal solution corresponds to the combination of the PV panel and batteries that guarantees a minimum LCC for total load satisfaction. The optimal results for different LLP levels are illustrated in Table 2.

For the highest LLP level (70%), the optimal solution corresponds to the lowest LCC equal to 1. 592 M €. It corresponds to 24 PV panels and 30 batteries. If only half of the required consumers demand is supplied (LLP = 50%), the optimal solution corresponds to the combination of 27 PV panels and 30 batteries for a total LCC equal to 1.594 M €. The load profile shape is highly modified for these LLP levels and essential load demand might be shed. If LLP is equal to 20%, the optimal solution corresponds to 46 PV panels and 34 batteries and LCC equals 1.840 M €. For

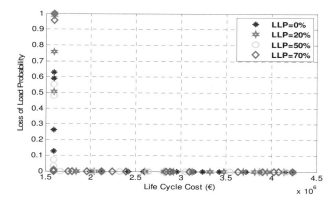

Figure 4.   Pareto front of LCC and LPSP for various LLP values.

Table 2.   Optimization results for different LLP levels.

LLP (%)	LCC (M€)	LPSP (%)	NPV	Nbat
70	1.592	0	24	30
50	1.594	0	27	30
20	1.840	0	46	34
0	2	0	30	37

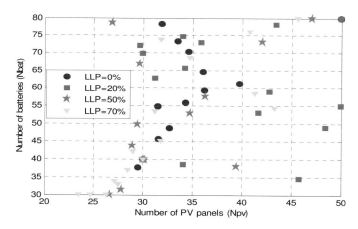

Figure 5.   Decision variables combination for different LLP and null LPSP.

the initial load profile the optimal combination is 30 PV panels and 37 batteries and LCC equal to 2 M€. The highest LCC corresponds to the original load profile. Applying modification on the daily load profile shape affects the required optimal combination of the decision variables and thus the economic criterion.

Figure 5 presents all the combinations of the decision variables Npv and Nbat that guarantee total load satisfaction (LPSP=0) for different rates of LLP.

It can be noticed that the highest number of batteries is required for the original load profile. Actually, the battery bank is solicited to insure consumers' satisfaction during the first and the second consumption peaks. It can be shown that increasing the load profile modification rate results in the decrease of the total required number for both the PV generator and the battery bank which explains economic cost minimization. The optimal combination choice depends basically on the consumer's and constructor's choice of optimization priority. It can change notably if a certain amount of power loss is acceptable and the priority is given to the minimization of the economic cost.

6 CONCLUSION

A bi-objective sizing optimization of a SAPV system was investigated. Both LCC and LPSP were considered as objective functions subject to the constraints related to the battery bank SOC. A study of the load profile management impact on the sizing optimization was carried out using the LLP rate. The optimization result was a Pareto front presenting different solutions for various values of LCC and LPSP. Introducing some changes on the load profile shape is beneficial in terms of the economic optimization criterion. Furthermore, the optimal combination varies significantly with the load profile shape. The rate of LPSP and LLP and the economic criterion choice depends on the consumers' preferences. The optimal combination depends on the priority that consumers give to each optimization criterion whether it is cost minimization with an acceptable LPSP and manageable load profile or total load satisfaction for a higher cost and no load profile changes tolerance.

ACKNOWLEDGMENT

This work was supported by the Tunisian Ministry of High Education and Research under the ERANETMED collaboration project "Energy and Water Systems Integration and Management" ID Number 044.

REFERENCES

Abdulaziz, N.I., Sulaiman, S.I., Shaari, S., Musirin, I. & Sopian, K. 2017. Optimal sizing of dtand-alone photovoltaic system by minimizing the loss of power supply probability. *Solar Energy (150)*: 220–228.

Bataineh, K. & Dalalah D. 2012. Optimal Configuration for Design of Stand-Alone PV System. *Smart Grid and Renewable Energy*: 139–147.

BP. 2019. *BP Energy Outlook*.

Cherif, H. & Belhadj, J. 2016. Synthesis and interest in hybridization of autonomous renewable multi-source systems under constraints of embodied energy GHG emissions life cycle cost and loss of power supply probability. *Proceedings of the International Conference on Recent Advances in Electrical systems*, Tunisia.

Deshmukh, M.K. & Deshmukh, S.S. 2008. Modeling of hybrid renewable energy systems. *Renewable and Sustainable Energy Reviews (12)*: 235–249.

Fezai, S. & Belhadj, J. 2016. Load profile impact on a stand-alone photovoltaic system. *7th International Renewable Energy Congress, Hammamet, 22–24 March 2016*, Tunisia.

International Energy Agency, 2018. *World Energy Outlook 2018 Executive Summary*.

Jain, S. & Jain, P.K. 2017. The rise of Renewable Energy implementation in South Africa. *Energy Procedia (143)*: 721–726.

Kartite, J. & Cherkaoui, M. 2016. Optimization of hybrid renewable energy power systems using Evolutionary algorithms. *Proceedings of the 5th International Conference on Systems and Control, Marrakech, 25–27 May 2016*. Morocco.

Lutton, E. 2005. *Artificial Darwinisme: an overview*.

Mahesh, A. & Sandhu, K. S. 2016. A genetic algorithm based improved optimal sizing strategy for solar-wind-battery hybrid system using energy filter algorithm. *Front. Energy*.

Okoye, C.O. & Solyali, O. 2017. Optimal sizing of stand-alone photovoltaic systems in residential buildings. *Energy*.

Ould Bilal, B., Sambou, V., Ndiaye, P.A., Kébé, C.M.F. & Ndongo. M. 2013. Study of the influence of load profile variation on the optimal sizing of a standalone hybrid PV/Wind/Battery/Diesel system. *Energy Procedia (36):* 1265–1275.

Pilai N.V. 2014. Loss of Load Probability of a Power Syste. *Journal of Fundamentals of Renewable Energy and Applications.*

Pillai, N. V. 2008. Loss of Load Probability of a Power System. *Munich Personal RePEc Archive.*

Renewable Energy Policy Network for the 21st Century, 2019. *Renewables 2019 Global Status Report*

Solar Power Europe, 2017. *Global Market Outlook For solar power 2017–2021.*

Tégani, I., Aboubou, A., Ayad, M.Y., Becherif, M., Saadi, R. & Kraa, O. 2014. Optimal sizing design and energy management of stand-alone photovoltaic/wind generator systems. *Energy Procedia (50):* 163–170.

Thiaux, Y., Seigneurbieux, J., Multon, B. & Ben Ahmed, H. 2010. Load profile impact on the gross energy requirement of stand-alone photovltaic systems. *Renewable Energy (35):* 602–613.

Wu, T., Xiao, Q., Wu, L., Zhang, J., Wang, M. 2011. Study and implementation on batteries charging method of Micro-Grid photovoltaic systems. *Smart Grid and Renewable Energy:* 324–329.

Zhou, W., Lou, C., Li, Z., Lu, L. & Yang, H. 2010. Current status of research on optimum sizing of stand-alone hybrid solar-wind power generation systems. *Applied Energy (87):* 380–389.

Innovative and Intelligent Technology-Based Services for Smart Environments – Ben Slama et al (eds)
© 2021 Taylor & Francis Group, London, ISBN 978-1-032-02030-3

Smart energy management based on the artificial neural network of a reverse osmosis desalination unit powered by renewable energy sources

A. Zgalmi, H. Cherif & J. Belhadj
Ecole Nationale Supérieure d'Ingénieurs de Tunis, Université de Tunis, Montfleury, Tunisie
Laboratoire des Systèmes Electriques, Ecole Nationale d'Ingénieurs de Tunis,
Université de Tunis El Manar, Tunis le Belvédère, Tunis, Tunisia

ABSTRACT: This paper aims to develop a Smart Energy Management System based on an artificial intelligence technique of a brackish water reverse osmosis desalination plant powered by stand-alone hybrid (PV/Wind) renewable sources. This system aims to meet the freshwater demand of an isolated community in a specific site of Tunisia's south. This study is characterized by a hydraulic storage in water tanks instead of electrochemical storage. For this purpose, a power management based on an Artificial Neural Network was developed to share power between three motor-pumps (High pressure pump, Low pressure pumps). The different components of the system (the motor-pumps, reverse osmosis, tanks, photovoltaic system and wind system) are defined with their energy models in order to size the system. This smart management is integrated in a dynamic simulator of the proposed system. The proposed strategy deals to maximize freshwater production taking interest in the available renewable energy.

1 INTRODUCTION

The Intergovernmental Panel on Climate Change (IPCC) has indicated that GHG emissions have increased by 1.6% per year, while carbon dioxide emissions have also increased by 1.9% per year over the last three decades. At the same time, water scarcity has become a major political and global concern considering both control and water sharing. 1.5 billion people do not have easy access to drinking water and some 80 countries suffer from water scarcity (Chen et al. 2016). In fact, water and electricity are fundamentally linked: energy is used to collect, treat and distribute water, while water is used to operate turbines and to cool the reactors. Water and energy management strategies should be coordinated due to the strong interdependence between vital water and energy resources. A promising solution to further enhancing the supply of fresh water is the desalination of brackish water (Elimelech & Phillip 2011; Greenlee et al. 2009; Ayati et al. 2019; Yang et al. 2020). Desalination requires much more energy than typical treatment of water. This energy can generate high financial costs for desalination operations and lead to significant carbon dioxide emissions (IEA 2015). Renewable energy technologies give a solution to meet the energy needs of desalination (Chen et al. 2019). The sizing, modeling and design of complex systems such as desalination with renewable energy systems are difficult, especially when electricity generated from renewable sources is used to produce fresh water for consumers of remote rural or urban areas (Ben Ali et al. 2018). To achieve more sustainable freshwater production, water and energy systems must be integrated into a single, efficient system. It is demonstrated in literature that much effort has been made in order to develop the optimal EMS for desalination units coupled with renewable energy. Kyriakarakos et al. (2017) showed an energy management technique based on Fuzzy Cognitive Maps to compare two cases of the capacity factor of the desalination unit operating only at full load studies to size the PV-battery system through optimization. Ben Ali et al. (2018) developed a Fuzzy Logic based Energy Management Strategy (FLEMS) to identify the power sharing between the different processes of SWRO desalination unit.

In this context, this paper aims to develop a management strategy (water-energy) based on evolutionary algorithm multi-criteria of a multi-source renewable energy system with hydraulic storage coupled to a pumping and desalination process. For this purpose, an ANN was developed. The neurons are connected with coefficients (weights) which constructs the neural network structure. There are input weights to transfer information and output weights to process information. For our case, with complex problems, MLP is the better model over the single-layer perception (Saud ALTobi 2019). The back-propagation learning algorithm is used to train the neural network with input and target pair patterns. The network takes the input and target values in the training data set and changes the value of the weighted links to reduce the difference between the output and target values using the training function, trainbr, of Levenberg-Marquardt.

Nomenclature	
GHG: Green House Gases	SWRO: Sea Water Reverse Osmosis
ANN: Artificial Neural Network	MLP: Multi-Layer Perception
MSE: Mean Squared Error	LP: Low pressure
HP: High pressure	BW: Brackish Water
BWRO: Brackish Water Reverse Osmosis	LPSP: Loss of Power Supply Probability
EMS: Energy Management System	

2 DESCRIPTION OF HYDRAULIC PROCESS POWERED BY HYBRID PV/WIND SYSTEM

The architecture of the desalination process coupled to hybrid PV/Wind system is shown in Figure 1. This system consists of an electrical part and a hydraulic part. The electrical part consists of a hybrid source which includes wind turbines, photovoltaic sources and converters (AC/DC) and (DC/DC). The intermittent power is distributed to the hydraulic part through a DC bus. The hydraulic part consists of three combinations of motor-pumps with different functions: the motor-pump LP1 operates at a fixed speed (50 Hz) and is used to pump the brackish water from the well and to store it in BW tank1. The motor-pump HP operates at variable speed, extracts brackish water from BW

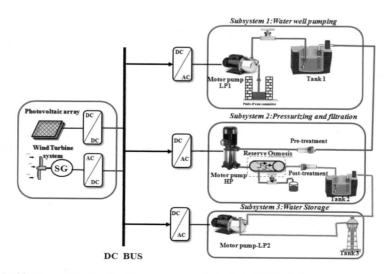

Figure 1.   Architecture of the desalination process coupled to hybrid PV-wind system.

tank1 and pressurizes it with the reverse osmosis module in order to produce permeated water. The motor-pump LP2 operates at a fixed speed (50 Hz) and moves permeated water up from tank2 to the water tower to feed consumers (Cherif et al. 2016). The profile load of the freshwater consumption, the wind speed, the solar irradiation and the ambient temperature have been recorded during one year with an hour acquisition period for the south of Tunisia (Cherif et al. 2016). The proposed system is characterized by hydraulic storage in water tanks instead of electrochemical storage.

3 SMART ENERGY MANAGEMENT BASED ON NEURAL NETWORK

3.1 Architecture of the selected ANNs

In this paper, authors used MLP Neural Network Architectures. This type of network has a multilayer feed forward structure and uses supervised training algorithms (Hornik et al. 1989). In this type of training, a set of inputs is presented to the network together with the expected output (Guliyev & Ismailov 2016).

To train the network, W_{ij}, the synaptic weights (which define the strength of a synaptic connection between two neurons, the presynaptic neuron i and the postsynaptic neuron j), are modified in a manner proportional to the error which is generated between the actual and expected output of the network. The model used in this work, shown in (Figure 2), has 24 neurons in the input layer which characterize: the hybrid power data P_{hyb}, the powers of three motor-pumps, the power limits, the sensors and the levels of three tanks. The authors used a single hidden layer with a Hydraulic Tangent activation function "TRANSIG" to activate the neurons of the hidden layer and the three neurons of the output layer.

The issue of the hidden layer number and the number of neurons in each layer remain a problem (Guliyev & Ismailov 2016; Arai 1993; Stathakis 2009; Ata 2015). In the literature, Cabrera et al. 2017; Jafar & Zilouchian 2001; and Barello et al. 2014 use two hidden layers of neurons in the ANN architecture to simulate a reserve osmosis process although some others have used up to 20 hidden layers (Barello et al. 2014)

3.2 Design and evaluation of the selected ANNs

A Parametric Sensitivity Analysis (PSA) has been developed to find the most appropriate ANN architecture. One of the major challenges when working with neural networks is the phenomenon of

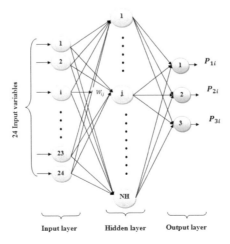

Figure 2. Schematic diagram of an ANN with 24 neurons in the input layer, one hidden layer of neurons (NH number of neurons) and three neurons in the output layer.

over fitting when the training algorithm runs too long. The number of hidden neurons can influence the stability of the neural network which is estimated by the error. The goal is to find the number of neurons in the hidden layer which minimizes the errors between the observed data and the data estimated by the ANN models.

The approach presented in (Figure 3) consists of dividing the data obtained by energy management based in the deterministic algorithm (Affi et al. 2018) into a number of disjointed subsets of similar size. The number of subsets depends on percentage of training, validation and testing which in this case is 80% training, 10% testing and 10% validation. For this purpose, the data of the subsets is divided into two blocks (training and validation block, testing block). The Bayesian back propagation algorithm (trainbr) was used in model training of all the MLP neural networks employed in this work. It is a network training function that updates the weights and biases according to the Levenberg-Marquardt rule. This algorithm was selected after a set of tests.

The PSA repeats the procedure and, using the former model and the data subsets, partial errors are determined. The final error of the sample is determined by calculating the arithmetic mean of the partial values. There is another technique that consists of using the same steps as the first one but replacing the MSE with a parameter called Training_R (the ratio between the outputs of neural networks and the desired outputs) to determine the number of neurons in the hidden layer. In our case, the second method is used as the easiest one to implant. This algorithm is implemented in the Matlab Neural Network Toolbox and it stops the training process when one of the criteria is fulfilled: a) The error obtained with the training data decreases and the error obtained with the validation data increases over six consecutive iterations, b) the iterations in the training process reach the value 1000 (Epochs= 1000), c) and a minimum MSE is reached (10^{-4}).

The number of neurons in the hidden layer can be a value between 50 and 150 (this interval is chosen following a set of tests to reduce the computation time). The goal is to find the number of neurons in the hidden layer that minimizes the errors between the desired data and the data estimated by the ANN model.

Figure 4 shows the results allowed about the number of neurons in the hidden layer in the ANN model developed from the BWRO desalination plant for different cases. We note that the result obtained for Nhopt= 106 is considered as the suitable solution for our work with training_R= 0,966.

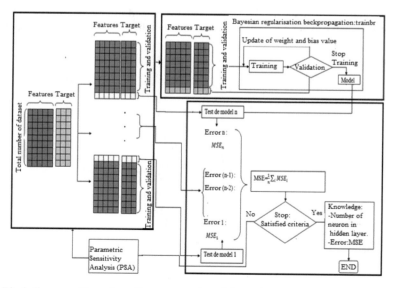

Figure 3.   Block diagram with a schematic representation of the methodology employed in the determination of the number of neurons in the hidden layer of the selected ANNs.

4 DYNAMIC SIMULATOR WITH SMART ENERGY MANAGEMENT

In order to study the performance of the presented system and to simulate the temporal evolution of variables, a dynamic simulator with the integrated smart power management was developed using MATLAB/SIMULINK. For this, data of wind speed, radiation and consumption are provided on a year with an acquisition every 60 minutes (hourly) (Zaibi et al. 2016). Meteorological data is linked to the project site (South Tunisia: Djerba).

The design of the dynamic simulator with the integrated smart power management (Figure 5) contains subsystems of both hydraulic and electrical process (Cherif et al. 2016). It is based on a wind turbine model, a photovoltaic generator model, a model of three-phase low-level pressure motor-pumps, a three-phase pump model for the storage of desalinated water at low pressure, a model of the unit of reverse osmosis desalination coupled to high pressure multi-stage three phase motor pump models, a model of brackish water storage tanks and desalinated water, a water tower model of fresh water and an intelligent energy management system based on ANN.

The desired hydraulic LPSP is included in the "LPSP" subsystem. The output of this block is the probability of dissatisfaction of demand expressed in percent (%).

5 RESULTS ANALYSIS

This section presents the simulation results of the proposed smart EMS of the hybrid PV-Wind system coupling water desalination process. The (Figure 6) shows the electrical power variation of three motor-pumps. At the beginning of this practical scenario (time=t1), tanks 1 and 2 are considered empty. During t1-t2, the EMS shares produced power between the two motor-pumps: motor-pump HP and motor-pump LP1, such that the motor-pump HP will receive the remaining power of the source after powering the motor-pump LP1.

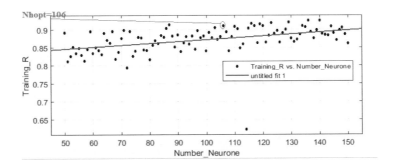

Figure 4. Results of the parametric sensitivity analysis.

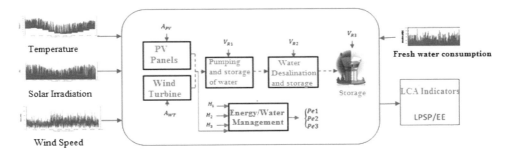

Figure 5. Dynamic simulator of the proposed system.

At t2, the EMS switched to share electrical power between the three motor-pumps, due to the increase in the electrical power emulated by the DC power source. Between t2-t3, the EMS continued to share power between the three motor-pumps. But at t3, the EMS decided to operate only the motor-pump LP2.

Figure 7 shows the levels of water in three tanks and the water consumption profile. At the beginning of this practical scenario (time=t1), tank2 is considered full. During t1-t2, the EMS shares produced power between the two motor-pumps: motor-pump HP and motor-pump LP2. At

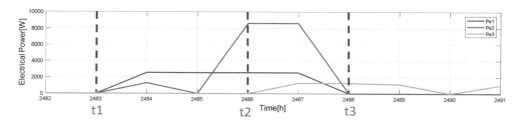

Figure 6.   Power sharing between the three motor-pumps based on NN power management.

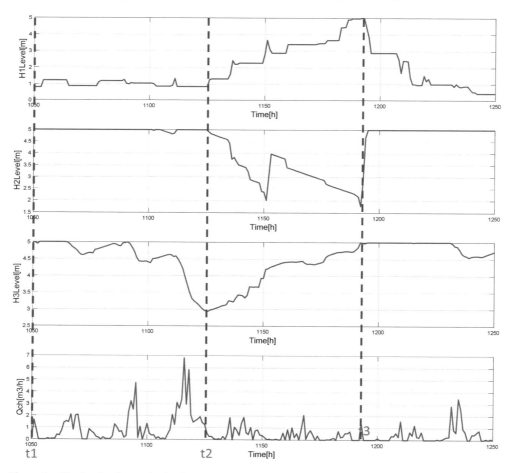

Figure 7.   The levels of water in the three tanks and the water consumption.

t2, the EMS switched to share electrical power between the three motor-pumps, due to the increase of water demand. During t2-t3, the EMS continued to share power between the three motor-pumps.

6 CONCLUSION

In this paper, a standalone PV/Wind BWRO desalination system has been presented. The purpose of this system is to provide the hydraulic storage in tanks when renewable energy is available instead of electrochemical storage. The authors developed a neural network energy management strategy for the system in order to share the power between three motor-pumps. Power sharing is based on ANNs who guarantee availability of systems over long periods with a good quality and quantity of water production. The major scientific challenge that this work contributes to rise is the development of EMS, which could work in real time for such a standalone system especially in remote communities. The objective of the developed EMS is to produce as much freshwater as much as possible taking advantage of the electric power availability offered by wind speed and radiation conditions. The ANN developed contains only one hidden layer with 106 neurons trained by the back-propagation algorithm with training function "trainbr." A dynamic simulator was developed containing the different elements of the system. The Neural Network decision presented encouraging performance in term of sharing the power, the quantity of fresh water produced and the energy efficiency.

ACKNOWLEDGMENT

This work was supported by the Tunisian Ministry of High Education and Research under the ERANETMED project "Energy and Water Systems Integration and Management" ID Number 044.

REFERENCES

Affi, S. Cherif, H. & Belhadj, J. 2018. Size optimization based on design of experiments of a water desalination processes supplied with hybrid PV/wind system with Battery storage. *Symposium de Genie Electrique (SGE), Nancy, France, 3–5 July 2018.*

Arai, M. 1993. Bounds on the number of hidden units in binary-valued three-layer neural networks. *Neural Networks* (6): 855–860.

Ata, R. 2015. Artificial neural networks applications in wind energy systems: a review, Renew. *Sustain. Energy Rev.* 49 (0): 534–562.

Ayati, E. Rahimi-Ahar, Z. Sadegh Hatamipour, M. & Ghalavand, Y. 2019. Water productivity enhancement in variable pressure humidification dehumidification (HDH) desalination systems using heat pump. *Applied Thermal Engineering* 160.

Barello, M. Manca, D. Patel, R. & Mujtaba, I. M. 2014. Neural network based correlation for estimating water permeability constant in RO desalination process under fouling. *Desalination* (345): 101–11.

Ben Ali, I. Turki, M. Belhadj, J. & Roboam, X. 2018. Optimized fuzzy rule-based energy management for a battery-less PV/wind-BWRO desalination system. *Energy* (159): 216–228.

Cabrera, P. Carta, J.A. González, J. & Melián, G. 2017. Artificial neural networks applied to manage the variable operation of a simple seawater reverse osmosis plant. *Desalination* (416): 140–156.

Chen, P.Y. Chen, S.T. Hsu, C.S. & Chen, C.C. 2016. Modeling the global relationships among economic growth, energy consumption and CO2 emissions. *Renewable and Sustainable Energy Reviews* (65): 420–431.

Chen, Ch. Jiang, Y. Ye, Z. Yang, Y. & Hou, Li'an. 2019. Sustainably integrating desalination with solar power to overcome future freshwater scarcity in China. *Global Energy Interconnection*: 098–113.

Cherif, H. & Belhadj, J. 2011. Large-scale time evaluation for energy estimation of stand-alone hybrid photovoltaic-wind system feeding a reverse osmosis desalination unit. *Energy* (36): 6058–6067.

Cherif, H. Champénois, G. & Belhadj, J. 2016. Environmental life cycle analysis of a water pumping and desalination process powered by intermittent renewable energy sources. *Renewable and Sustainable Energy Reviews* (59): 1504–1513.

Elimelech, M & Phillip, W. A. 2011. The Future of Seawater Desalination. *Energy, Technology, and the Environment. Science* (333): 712–717.

Greenlee, L. F. Lawler, D. F. Freeman, B. D. Marrot, B. & Moulin, P. 2009. Reverse osmosis desalination. Water sources, technology, and today's challenges. *Water Research* 43(9): 2317–2348.

Guliyev, N. J. & Ismailov, V. E. 2016. A Single Hidden Layer Feed forward Network with Only One Neuron in the Hidden Layer Can Approximate Any Univariate Function. *Neural Computation* (28): 1289–1304.

Hornik, K. Stinchcombe, M. & White, H. 1989. Multilayer feed forward networks are universal approximators. *Neural Networks* (2), 359–366.

International Energy Agency, *World outlook Energy* 2015.

Jafar, M. & Zilouchian, A. 2001. Intelligent control systems using soft computing methodologies, Ch. Application of Soft Computing for Desalination Technology, CRC. *Press, London*: 317–349.

Kyriakarakos, G. Anastasios, I. Dounis, Konstantinos, G. Arvanitis, Papadakis, G. 2017. Design of a Fuzzy Cognitive Maps variable-load energy management system for autonomous PV-reverse osmosis desalination systems. A simulation survey. *Applied Energy* 187: 575–584.

Saud ALTobi, M. Ali. Bevan, G. Wallace, P. Harrison, D. & Ramachandran, K.P. 2019. Fault diagnosis of a centrifugal pump using MLP-GABP and SVM with CWT. *Engineering Science and Technology, an International Journal* 22: 854–861.

Stathakis, D. 2009. How many hidden layers and nodes? *International Journal of Remote Sensing* (30): 2133–2147.

Yang, H. Fu, M. Zhan, Z. Wang, R. & Jiang, Y. 2020. Study on combined freezing-based desalination processes with microwave treatment. *Desalination* 475: 114–201.

Zaibi, M. Champenois, G. Roboam, X. Belhadj, J. & Sareni, B. 2016. Smart power management of a hybrid photovoltaic/wind stand-alone system coupling battery storage and hydraulic network. *Mathematics and Computers in Simulation*: 1–22.

Innovative and Intelligent Technology-Based Services for Smart
Environments – Ben Slama et al (eds)
© *2021 Taylor & Francis Group, London, ISBN 978-1-032-02030-3*

Hybrid concentrated solar power plant and biomass power plant

A. Jemili, S. Ferchichi, E. Znouda & C. Bouden
École Nationale d'Ingénieurs de Tunis, Université Tunis el Manar, Tunis, Tunisia

ABSTRACT: A novel hybrid Concentrated Solar Power (CSP) and Biomass power plant is installed at Ecole Nationale d'Ingénieurs de Tunis (ENIT). It is the first CSP plant in Tunisia, using Parabolic Trough Collectors with Direct Steam Generation to provide 60 KW of electricity through an Organic Rankine Cycle Turbine. Additionally, biomass anaerobic digestion is used as auxiliary energy to back up the CSP system. This paper includes details about the sizing of the biomass system. Hybrid operation mode (solar field and boiler) test results are presented and assessed, highlighting the potential of implementing CSP and biomass hybridization in Tunisia.
Keywords: Biomass system, Hybrid Power Plant, Biogas production, Hybrid Boiler

1 INTRODUCTION

The energy transition in Tunisia is standing at an important crossroads. The country faces a rise in energy demand driven by continued population and economic growth. This leads to a challenging situation for a country that relies on energy imports. On one hand, around 97% of the total electricity is generated by one source, which is natural gas (ANME & GIZ 2014). On the other hand, the meteorological and geological conditions in Tunisia reveal a high potential of renewable energies, especially solar. The country has very good solar radiation potential, which is ranging between 1800 kWh/m^2 per year up in the North and 2600kWh/m^2 per year in the South (Balghouthi et al. 2016). Tunisia must enhance its energy security and meet long-term social and economic goals by achieving better efficiency and greater reliance on renewables. The country has set out to increase and diversify its use of renewables. This has involved establishing a new target whereby 30% of the country's total electricity consumption which would come from renewable sources by the year 2030 according to the Tunisian Solar Plan (PST)(2015).

As a renewable source, solar energy is intermittent and can cause disruption of the electric energy supply. Thus, it is necessary to use storage or backup systems to ensure a balance between electricity demand and supply (Oliveira 2013). The flexibility of bioenergy makes it a suitable solution to overcome the problem of intermittency (Angrisani et al. 2013). Biomass and CSP Hybridization is not a new concept. Nevertheless, during the past seven years, the interest in this combined system has increased. Several research studies have been carried out in this field. Research focus was mainly on (i) plant efficiency, (ii) the LCOE of the produced electricity and (iii) environmental impacts (Cheng 2010; Nixan et al. 2012).

In this context, a mini hybrid concentrated solar power (CSP) plant has been implemented, commissioned and tested at Ecole Nationale d'Ingénieurs de Tunis (ENIT), in the framework of REELCOOP (Renewable Electricity COOPeration) project (Krüger 2015). The plant relies on Parabolic Trough Collector (PTC) technology with Direct Steam Generation (DSG). In order to back up the solar field, auxiliary energy is provided by a steam boiler using Biogas as fuel (Willwerth et al. 2017). The vapor generated in the solar field and/or boiler is used to drive an Organic Rankine Cycle (ORC). The biogas is produced through anaerobic digestion process of food waste from the university canteen. A previous work (Jmili 2015) has shown that the potential yield of Biogas from the canteen waste is promising. Biomass system sizing results are presented as well as the overall commissioning and testing results of the plant in hybrid operation mode (solar and boiler). Both sets of results are thoroughly discussed.

2 PLANT LAYOUT

The small-scale power plant of 60 kW with DSG has been erected at the premises of ENIT. The developed prototype is based on the hybridization of two different renewable resources, solar and biomass, for electricity generation in small scale (Figure 1).

In the biomass system, the biogas is produced by anaerobic digestion using the organic waste of the university restaurant. Anaerobic digestion is a biological process. During this process, microorganisms (bacteria) break down the organic matter in the absence of oxygen. This degradation leads to the formation of a biogas rich in methane (CH_4)(Bajpai 2017). A simplified scheme of the biogas prototype is represented in the following picture (Figure 2). The process starts with crushing

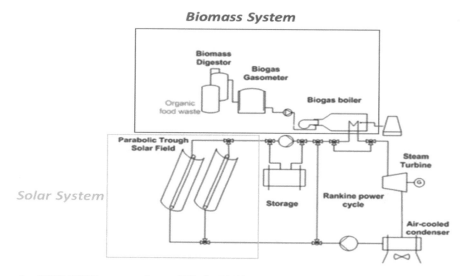

Figure 1. REELCOOP prototype layout (Oliveira 2013).

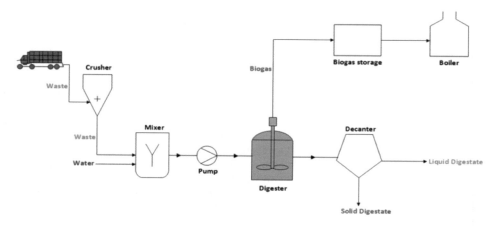

Figure 2. Schematic of biogas production process.

the feedstock of the organic matter (to reduce particle size) and afterwards the obtained substrate is mixed with water into the mixer tank. The digester is fed daily through a pump and produces digestate. It is a one stage wet continuous digester, known as a complete stirred tank reactor. It is mechanically stirred. A slowly turning agitator ensures that the substrate is optimally mixed and that biogas is released. The anaerobic digestion process requires a constant temperature of 37°C (Tabatabaei & Ghanavati 2018), thus the digester is insolated and heated by means of solar thermal tubes evacuated in order to achieve a temperature of 37°C for the anaerobic digestion process and to compensate the heat losses. A heating system using solar collectors was developed to regulate the temperature during the digestion process. The digestate is dewatered, ensuring the production of processed water and press cakes using a decanter. To improve the demand response, a gasometer is used to store the produced and dry biogas before being used in the steam boiler.

3 SIZING OF THE BIOMASS SYETEM

As an auxiliary heat source, a steam boiler was chosen for this plant. The boiler is driven by biogas. The output steam pressure and temperature required to drive the ORC are 8.9 bars and 175°C respectively.

When sizing the boiler, the steam flow rate was calculated using the following equation (1) and using the data in Table 1.

$$\dot{m} = P_u/(h2 - h1) \tag{1}$$

The steam flow rate obtained is equal to $0.18\,\text{Kgs}^{-1}$

The boiler inlet thermal power is calculated using equation (2), presented below, and is considering a boiler efficiency of 85%.

$$P_i = \frac{P_u}{Boiler\ efficiency} \tag{2}$$

The biogas flow rate needed to operate the boiler is calculated based on the boiler inlet thermal power and the biogas Lower Heating Value (LHV). The results obtained are presented in Table 2.

It has been decided to operate the boiler 8 hours in winter and 5 hours in summer in order to assure the continuous operation of the power plant. Daily Biogas volumes of 480 m³ and 786 m³ are needed to run the boiler for 5 hours and 8 hours respectively.

The substrate used in this research work was collected from the university restaurant waste. It is mainly composed of fruits and vegetables waste as well as waste from cooked food. A detailed analysis was carried out in order to identify the characteristics of this substrate. Total Solids fraction

Table 1. Input data for the boiler sizing.

Input data	
Rated steam temperature	175°C
Feed water temperature	80°C
Output thermal power	450 kW
Steam enthalpy	2772.13 kJ/kg
Water enthalpy	334.96 kJ/kg

Table 2. The biogas flow rate needed.

Boiler inlet thermal power	528 kW
LHV Biogas (LHV)	5.5 kWh/m³
Biogas flow rate ($\dot{V}$)	96 m³/h

(TS) and volatile solids fractions (VS) were determined according to the standard methods 2540B and 2540E (Libetrau et al. 2016). The results are presented in Table 3. It should be noted that the presented values can vary daily due to change in canteen organic waste composition.

The daily quantity of the volatile solids (VS) in the substrate can be estimated through the specific gas yield, the daily estimated biogas production and the canteen organic waste properties (Soares et al. 2016) (Table 4).

$$\dot{m}_{vs} = \frac{\dot{V}_{biogas}}{Biogas\ yield} \qquad (3)$$

When sizing the fermentation tank (digester), the most important parameter that should be defined is the volume of the tank. Equation 4 was used to calculate the reactor volume (Arango-Osorio 2019; Tucki et al. 2015).

$$V_{reactor} = V_{sub} \times HRT \times 1.25 \qquad (4)$$

Calculation results have shown that a digester having a volume of to 670 m^3 is needed to operate the boiler. Besides, a biogas storage system is required to store the daily production and to be used at appropriate working hours. The burner has been chosen to operate both by using Biogas or natural gas depending on the availability.

The amounts of gas needed to operate the steam boiler with biogas or natural gas are presented in the previous graph (Figure 3). However, due to the constraints listed above and because of budget

Table 3. Canteen organic waste properties.

Organic waste properties	
Density of organic waste [OW]	0.7 Tons/m^3
Total solids fraction [%TS]	21%
Volatile solids fraction [%VS]	63%
Biogas yield	0.615 m^3/kg of VS

Table 4. The daily quantity of the organic waste.

The daily input of (VS) in the substrate ($\dot{m}_{vs}$)	1278 kg/day of VS
Daily Organic waste quantity (Mass)	10 Tons/day

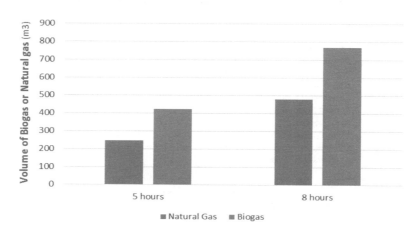

Figure 3. Volume of biogas and natural gas needed to run the boiler.

limitations, the boiler is operated using natural gas only and a digester of 3m^3 was manufactured for demonstration purposes.

4 OVERALL COMMISSIONING AND TESTING OF THE PLANT IN HYBRID OPERATION MODE

The tests of the boiler were carried out with natural gas and the organic waste anaerobic digestion performance was independently assessed. The boiler was successfully commissioned in July 2018 (Figure 4) (Krüger et al. 2017).

The first test of the installation with the boiler took place in October 2018, and the ORC was operated continuously using the steam produced by the boiler and then the solar field and boiler were operated in full load. During the operation, the collectors were focused for a restricted period due to hazy sky. Thus, the outlet temperature was not high enough to operate the ORC. The solar collector and the boiler operated simultaneously for about 40 minutes, and following that, the operation was carried out with the boiler only. The measurements during this test day are presented in Figure 4. The electric power was generated in the ORC in this discontinuous way due to the startup, shutdown and testing of the ORC, as well as the regulation of the control system. The electrical power produced reached a power ranging between 60 kW and 70 kW (Figure 5).

The plant was operated again in June 2019. These experimental tests included hybrid operation of the plant. The solar field and the boiler were successfully operated in parallel. The plant was operated in solar only mode during the first half of the day. The electrical power generated did exceed 40 kW during the morning. This low output was principally due to the low mass flow in the solar field. During the second half of the testing day, the plant was operated in hybrid mode. The solar field was started first and then the boiler. The electrical power produced ranged between 60 kW and 65 kW (Figure 6).

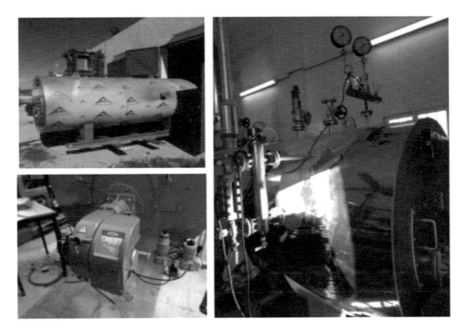

Figure 4. View of the steam boiler.

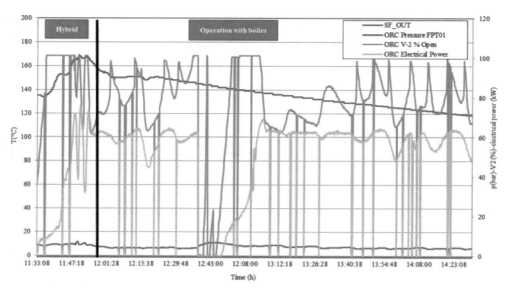

Figure 5. The performance of the test facility during one testing day in October (19/10/2018).

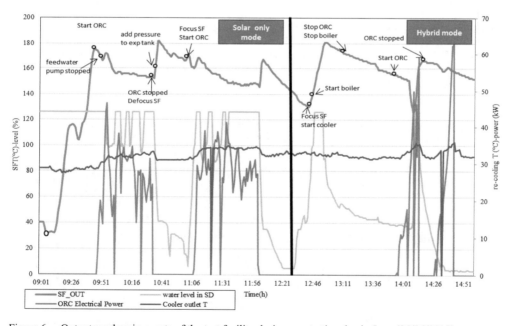

Figure 6. Outputs and main events of the test facility during one testing day in June (25/06/2019).

5 CONCLUSION

In the framework of REELCOOP project, a novel small-scale hybrid CSP and Biomass power plant was successfully commissioned at ENIT. The tests in hybrid mode allowed the ORC to operate in full load and the maximum power generated reached 70 kW. A mini digester of 3 m³ was installed for demonstration purposes. The Biogas prototype was commissioned in March 2018. A leakage test was performed first. During the test, gas losses in the digester were detected. After solving this problem, the biogas unit was successfully commissioned on August 2019. The experimental tests on biogas production will be carried out in order to assess the performance of the biomass system.

ACKNOWLEDGEMENTS

This work has been financed by FP7-REELCOOP project, which receives funding from the European Union Seventh Framework Programme (FP7/2007-2013), under grant agreement n°608466. The authors wish to thank all the members of REELCOOP work team.

REFERENCES

Agence Nationale de Maîtrise de l'Energie. 2015. Plan Solaire Tunisien.
Angrisani, G. et al. 2013. Development of a new concept solar-biomass cogeneration system. *Energy Conversion and management*: 75: p. 552–560.
ANME and GIZ. 2014. Stratégie nationale de maitrise de l'énergie Objectifs : moyens et enjeux.
Arango-Osorio, S. 2019. Methodology for the design and economic assessment of anaerobic digestion plants to produce energy and biofertilizer from livestock waste. *Science of the total environment*: 685:1169–1180.
Bajpai, P. March 2017. Basics of anaerobic digestion process. *Chapter. In book: Anaerobic technology in pulp and paper industry*: pp. 7–12.
Balghouthi, M. Trabelsi, M. BenAmmar, M. Bel Haj Ali, A. Guizani, A. 2016. Potential of concentrating solar power (CSP) technology in Tunisia and the possibility of integration with Europe. *Renewable and Sustainable Energy reviews*: Volume.56(C): pp. 1227–1248.
Cheng, J. 2010. Biomass to Renewable Energy Processes. *Taylor and Francis Group.*
Jmili, A. 2015. Experimental study of a biogas system using 1 m^3 digester. *1st International Conference Green Business and Sustainable Development*: Hammamet Tunisia.
Krüger, D. et al. Nov. 2015. Deliverable 4.2 – REELCOOP prototype 3 (CSP).
Krüger, D, Willwerth, L, and Dathe, S. 2017. Research Cooperation in Renewable Energy Technologies for Electricity Generation REELCOOP Report on commissioning and testing. *German Aerospace Center DLR.*
Libetrau, J, Pfeiffer, D, Thran, D. 2016. Collection of methods for biogas. *Series of the finding programme "Biomass energy use":* Volume 7.
Nixan, J.D. et al. 2012. The feasibility of hybrid solar-biomass power plants in India. *Energy*:46: 541–554.
Oliveira. A. 2013. Research Cooperation in Renewable Energy Technologies for Electricity Generation Part B, *University of Porto – Faculty of Engineering.*
Soares, J. et al. 2016. Deliverable 7.1: Prototype 3 installation and commissioning report.
Tabatabaei, M. and Ghanavati, H. 2018. Biogas: Fundamentals, Process, and Operation. *Springer*
Tucki, K., Klimkiewicz, M, Piatkowski, P. 2015. Design of digester Tank. *Teka Commission of motirization and energetics in agriculture*: vol.15: 83–88.
Willwerth, L., Rodriguez, M., Rojas, E., Ben Cheikh, R., Ferchichi, S., Jmili, A., Baba, A., Soares, J., Parise, F., Weinzierl, B., Kruger, D. September 2017. Commissioning and test of a mini CSP plant. *Conference Solar Paces, Santiago.*

NOTATIONS

h_2: Steam enthalpy (kJ/kg)
h_1: Water enthalpy (kJ/kg)
HRT: Hydraulic retention time (Days)
$\dot{m}$: steam flow rate (kg/s)
$\dot{m}_{vs}$: The daily input of volatile solids in the substrate (kg/day of VS)
P_u: Useful power (kW)
P_i: Inlet power (kW)
$\dot{V}_{biogas}$: Biogas daily production (m^3/day)
$V_{reactor}$: Reactor volume (m^3)
V_{sub}: The inlet volumetric flow of the substrate (m^3/day)

Innovative and Intelligent Technology-Based Services for Smart Environments – Ben Slama et al (eds)
© 2021 Taylor & Francis Group, London, ISBN 978-1-032-02030-3

Literature review on monitoring systems for photovoltaic plants

H. Messaoudi Abid & I. Slama-Belkhodja
ENIT, Université de Tunis El Manar, Tunis, Tunisia

H. Ben Attia Sethom
ENICarthage, Université de Carthage, Tunis, Tunisia

A. Bennani-Ben Abdelghani
INSAT, Université de Carthage, Tunis, Tunisia

H. Sammoud
APPCON Technologies, Ariana, Tunisia

ABSTRACT: Photovoltaic power generation systems have been globally installed in recent years and they are today widely adopted as one of the most inherent and cost-competitive renewable energy sources. Consequently, photovoltaic plant monitoring is becoming a crucial task to guaranteeing a stable and reliable operation in addition to effective cost management. This paper presents a detailed review of different monitoring systems reported in the literature, which strongly depend on the photovoltaic installation topology and output power range. This involves various monitoring techniques and the monitoring system concerns, opportunities and challenges.

1 INTRODUCTION

As a consequence of the drastic shrinking of fossil fuel resources on one hand, and the continuous evolution in energy demand due to the population growth and the significant industrial development worldwide on the other hand, many efforts were made to find alternative energy resources. Solar energy is one of the cleanest, most sustainable and most abundantly explored energy sources (Liu et al. 2017). Hence, PV systems are widely installed around the world. The research conducted on the PV systems do not focus only on increasing the installed capacity levels, but also on enhancing system performance and efficiency. Indeed, the PV plant performances are strongly affected by the weather conditions and the location where it's installed. In some locations, the panel soiling, for example, can cause about 70% power loss (Woyte et al. 2013) and partial shading around 10 to 20% (Hanson et al. 2016). Accordingly, continuous monitoring of the installation and healthy operation is a noteworthy concern to keep track of the PV plant energy production and thus to help in maintaining and improving its performance.

Various PV monitoring systems have been reported in the literature (Madeti & Singh 2017; Rahman et al. 2018). Blaesser (1997) proposed one of the first developed monitoring systems that allow collecting, analyzing and displaying PV operational data. However, the price of the data acquisition equipment at that time is about 10% higher than the overall system cost (Fuentes et al. 2014). One of the first low cost developed hardware designs of PV monitoring systems is presented in Mukaro et al. (1998). It consists of an 8-bit microcontroller for data acquisition and processing and a serial EEPROM for data storage. Another attempt to design an integrated PV monitoring system is reported in Koutroulis & Kalaitzakis (2003). This proposed data acquisition system's main limitations are the high cost and its dependence on the computer.

DOI 10.1201/9781003181545-28

Furthermore, different types of commercial monitoring systems for small and large PV plants have been proposed by multiple well-known companies (Rahman et al. 2018) such as the "Outback Power Flexware FN-DC Flexnet Advanced DC System Monitor" (FLEXnet DC-OutBack Power Inc) and "TED Pro Home Energy Monitor" (TED Pro). These commercial solutions provide web-based services, which allow the users to observe the PV output behavior. Some of these applications provide more useful services like alarms or notifications as text messages or e-mails in case of problems. However, these commercial monitoring systems are expensive, energy-consuming, require high storage capacities and do not allow the user to add new functionalities (Rahman et al. 2018). Therefore, and due to the availability of large and open source tools and hardware in addition to the development of new suitable algorithms and the increase of the sensors and data acquisition equipment (Tyagi et al. 2018), many researchers have introduced a more accurate, more cost-effective and more autonomous PV monitoring systems. The developed monitoring systems are based on various approaches including miscellaneous methods and communication technologies.

In this paper, a literature review of different monitoring techniques proposed for PV plants, the challenges and concerns related to their design and an overview of different monitoring systems reported in the literature are presented. The paper focuses on the multiple employed monitoring methods as well as the adopted technologies and topologies associated with them.

The paper is organized as follows: section 2 is dedicated to the different challenges and research trends related to the PV plant monitoring task. A detailed review of the most used monitoring techniques is depicted in section 3. In section 4, some conclusions about the presented studies are drawn.

2 PV MONITORING ISSUES AND RESEARCH INTENTIONS

The main focus of the monitoring task is to guarantee its sustainability by ensuring an optimal performance operation with acceptable returns on the investment performed by the project owner. This can be achieved by continuous monitoring of PV power plant performance. Indeed, a PV monitoring system aims to confirm the proper operation of the PV system, evaluate the efficiency of its components and locate the faulty devices when the desired performances are not reached and allow a remote inspection and diagnosis in place of human workforce involvement, which is dangerous, less accurate and time-consuming. This alternative minimizes the diagnosis and maintenance costs.

Continuous PV monitoring is not conceptually complex, but it should consider a number of constraints such as the size of the PV plant, the installation location nature, the climatic conditions, the sensitivity of the data generated by the PV system that is strongly influenced by external factors like the weather and the dirt (Ejgar & Momin 2017; Madeti & Singh 2017). Despite its importance, the PV monitoring project fails to reach its objective (Ejgar & Momin 2017) in many cases. Several reasons for this failure have been revealed in the past few years. For example, in many PV installation projects, the main intention is the establishment of the PV system itself rather than ensuring its efficiency and proper operation. Moreover, the monitoring is chronologically planed as the last task in the project. Thus, it can be sacrificed in case of time restrictions or budget excesses. Nonqualified or improperly trained personnel, which are in charge of monitoring, is also one of the serious causes of this failure. Besides this, the climatic conditions and the nature of the site where the PV system is installed can make monitoring more difficult and lead to its failure.

In the following section, a literature review of various proposed PV monitoring systems, which considers the above-cited challenges and intentions, is presented.

3 MONITORING SYSTEMS: OVERVIEW

The elementary architecture of the PV monitoring system is shown in Figure 1. It includes five main functional units, namely: a data acquisition unit gathering all used sensors and their conditioning

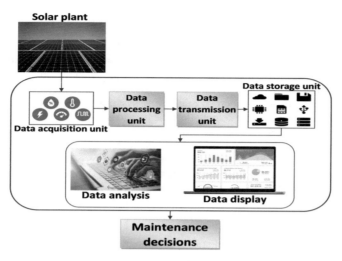

Figure 1.   PV monitoring system general architecture.

circuits, a data processing unit constituting the main brain of the system where the data is gathered and processed, a data storage unit where the collected data is stored, a data transmission unit and a data analysis and display unit, which are both responsible for the analysis, the diagnosis and the presentation of the data. The tasks ensured by each unit depend on the size, the configuration, and the connection topology of the PV plant. They depend also on the location and the weather conditions.

In recent years, numerous cost-effective PV plant monitoring systems have been reported in the literature. These systems differ from the point of view of functional unit tasks and used technologies. Adhya et al. proposed a microcontroller-based PV monitoring system in Adhya et al., (2016). In the proposed design, voltage transducers, current transducers, voltage dividers and shunt resistors are used to measure different currents, voltage quantities and an LM35 is used for temperature measurement. The data processing unit's main brain is the PIC18F46K22 microcontroller. The RS-232 serial port ensures data transmission to a personal computer. The communication between the data processing unit and the server is ensured through a GPRS module. Chouder et al. presented a LabVIEW-based real-time monitoring tool for a grid-connected PV system in Chouder et al. (2012). This system allows measuring the temperature, the solar radiation, the voltages and the currents of both DC and AC sides using thermocouples, pyranometers, voltage dividers and Hall-effect current sensors. The collected data is processed under LabVIEW and displayed on a visualization interface. The data transfer is ensured by a GPIB bus. Ahmad Rivai et al. (2020) proposed a low-cost system for the monitoring of PV array efficiency and MPPT algorithm performance. The electrical parameters, namely the DC voltage and current, the solar radiation and the temperature were measured using a voltage divider, a shunt resistor, a pyranometer LI-200 and the LM35 sensor, respectively. For data processing, analysis and display, an ATMega8535 microcontroller connected to a personal computer through a master/slave model was adopted. A low cost resistive I-V tracer is also used to determine the PV array I-V curve, which can be displayed later on the microcontroller LCD screen. In Caruso et al. (2015), an ATmega 328P-PU microcontroller-based PV monitoring system allows acquiring the DC current and voltage of the PV system, processing the related data and transmitting it using a set of DC smart meters wirelessly interconnected to a central controller with a radiofrequency of about 315 MHz. The collected data is stored on an SD memory card and displayed on a user interface that communicates with a central controller via Bluetooth. The PV monitoring system depicted in Hashmi et al. (2020) is based on a10-bit Atmega 328 microcontroller of the Arduino UNO R3. This monitoring system is developed to monitor the DC current and voltage provided by the PV module using the current sensor ACS712 and a voltage sensor. The measurements are transferred to the Ardunio board, which is connected to a personal computer through USB cable. Li Bian *et al.* presented a multifunctional data acquisition system in Bian et al. (2012) for real-time electrical and metrological parameter monitoring. For data acquisition, the NI cRIO platform with different sensors like LEM voltage and current sensors, PT100 temperature sensor, and pyranometer have been used. LabVIEW-based software is designed for processing, analyzing and displaying different collected data. The acquired data is transferred to a personal computer via Ethernet, where it is displayed and stored. In (Ranhotigamage & Mukhopadhyay 2011), the PV current and voltage are measured using a shunt resistor and a voltage divider, respectively. The NTC thermistor B57164K472J measures the temperature and a photodiode PDB-C139 measures the radiation. The measured data is processed using a microcontroller and is transferred to the user via the Zigbee communication protocol. A TinyOS-based remote intelligence system has been introduced in Jihua & Wang (2014). In this case, the data is collected using wireless sensor networks that include sink nodes and sensor nodes. Then, the data is processing in an ARM gateway. Another system design for temperature monitoring has been proposed in (Gad & Gad 2015). The LM35 is used to measure the temperature. The measured data is stored daily onto an SD memory card and processed in an ATMega2560 microcontroller. Maulik et al. introduced a new design of PV monitoring systems in Vyas et al. (2016), where the data is acquired using a voltage sensor, current sensors, a temperature sensor and a light intensity sensor. These measurements are then transferred to a DAQ card to be converted to digital quantities. After that, it's interfaced with LabVIEW for further analysis and display. Adhya et al. proposed (Adhya et al. 2016) an IoT-based remote monitoring system. This system allows measuring the battery voltage and the PV voltage using a voltage divider circuit, the battery current and the PV current using a shunt with differential amplifiers, the grid voltage using a potential transformer, the grid current using a current transformer, the solar radiation using a unit solar cell with a precision amplifier and the temperature using the LM35 sensor. The collected data is processed in a PIC18F46K22 microcontroller and locally stored onto a Micro SD memory card. The data is then transmitted to a remote server through the mobile radio network employing a GPRS module. In Begum et al. (2016), an O&M system based on predictive analytics and SCADA is introduced. In this case, the collected analog data acquired by different sensors like the LTC2990 – Quad I2C are converted to digital quantities by an ARM processor ADC. Google Cloud's BigTable is adapted for storing the big amount of collected data along with IoT devices.

4 CONCLUSION

In this paper, the main influencing issues and research trends, which have to be considered and dealt with in any PV monitoring project, are introduced. Then, a comprehensive literature review of different proposed PV monitoring systems and techniques is presented. Indeed, earlier proposed monitoring systems were wired and involved complex and expensive technologies. Then, along with the technological development and the widespread open-source hardware and software, the intention in PV monitoring systems was introduced as making low cost, wireless and easy to use designs. PV monitoring system architecture is based on five different units, namely the acquisition unit, the processing unit, the transmission unit, the storage unit and the analysis and display unit. Thus, the presented literature review is focused on the used technologies, components and topologies adopted for each functional unit of these systems.

REFERENCES

Adhya, S. et al. 2016. An IoT based smart solar photovoltaic remote monitoring and control unit. *2nd international conference on control, instrumentation, energy and communication (CIEC)*: 432–436.
Begum, S. et al. 2016. A comparative study on improving the performance of solar power plants through IOT and predictive data analytics. *International Conference on Electrical, Electronics, Communication, Computer and Optimization Techniques (ICEECCOT)*: 89–91.

Bian, L. et al. 2012. A multifunctional data acquisition system for photovoltaic plants. *International Conference on Systems and Informatics (ICSAI2012)*: 598–602.

Blaesser, G. 1997. PV system measurements and monitoring the European experience. *Solar energy materials and solar cells*, 47(1–4): 167–176.

Caruso, M. et al. 2015. A low-cost, real-time monitoring system for PV plants based on ATmega 328P-PU microcontroller. *IEEE International Telecommunications Energy Conference (INTELEC)*: 1–5.

Chouder, A. et al. 2012. Monitoring, modelling and simulation of PV systems using LabVIEW. *Solar Energy*, 91: 337–349.

Ejgar, M. & Momin, B. 2017. Solar plant monitoring system: A review. *International Conference on Computing Methodologies and Communication*: 1142–1144.

FLEXnet DC-OutBack Power Inc, http://outbackpower.com/products/system-management/flexnet-dc, last accessed 2020/01/13.

Fuentes, M. et al. 2014. Design of an accurate, low-cost autonomous data logger for PV system monitoring using Arduino that complies with IEC standards. *Solar Energy Materials and Solar Cells*, 130: 529–543.

Gad, H. E. & Gad, H. E. 2015. Development of a new temperature data acquisition system for solar energy applications. *Renewable energy*, 74: 337–343.

Hanson, A. J. et al. 2016. Partial-shading assessment of photovoltaic installations via module-level monitoring. *IEEE Journal of Photovoltaics*, 4(6): 1618–1624.

Hashmi, G. et al. 2020. Portable solar panel efficiency measurement system. *SN Applied Sciences*, 2(1): 56.

Jihua, Y. & Wang, W. 2014. Research and design of solar photovoltaic power generation monitoring system based on TinyOS. *9th International Conference on Computer Science and Education*: 1020–1023.

Koutroulis, E. & Kalaitzakis, K. 2003. Development of an integrated data-acquisition system for renewable energy sources systems monitoring. *Renewable Energy*, 28(1): 139–152.

Liu, G. et al. 2017. Architecture and experiment of remote monitoring and operation management for multiple scales of solar power plants. *IEEE 2nd Advanced Information Technology, Electronic and Automation Control Conference (IAEAC)*: 2489–2495.

Madeti, S. R. & Singh, S. N. 2017. Monitoring system for photovoltaic plants: A review. *Renewable and Sustainable Energy Reviews*, 67: 1180–1207.

Mukaro, R. et al. 1998. First performance analysis of a silicon-cell microcontroller-based solar radiation monitoring system. *Solar Energy*, 63(5): 313–321.

Rahman, M. M. et al. 2018. Global modern monitoring systems for PV based power generation: A review. *Renewable and Sustainable Energy Reviews*, 82: 4142–4158.

Ranhotigamage, C. & Mukhopadhyay, S. C. 2011. Field trials and performance monitoring of distributed solar panels using a low-cost wireless sensors network for domestic applications. *IEEE Sensors Journal*, 11(10): 2583–2590.

Rivai, A. et al. 2020. Analysis of Photovoltaic String Failure and Health Monitoring with Module Fault Identification. *Energies*, 13(1): 100.

TED Pro Home is latest technology in home energy management, https://www.theenergydetective.com/tedprohome.html, last accessed 2020/01/13.

Tyagi, A. et al. 2018. Advance Monitoring of Electrical and Environmental Parameters of PV System: A Review. *International Conference on Sustainable Energy, Electronics, and Computing Systems (SEEMS)*: 1–5.

Vyas, M. et al. 2016. Real Time Data Monitoring of PV Solar cell using LabVIEW. *Int. J. Curr. Eng. Int. J. Curr. Eng. Technol*, 6(6): 2218–2221.

Woyte, A. et al. 2013. Monitoring of photovoltaic systems: good practices and systematic analysis. *Proc. 28th European Photovoltaic Solar Energy Conference:* 3686–3694.

Innovative and Intelligent Technology-Based Services for Smart
Environments – Ben Slama et al (eds)
© 2021 Taylor & Francis Group, London, ISBN 978-1-032-02030-3

Control of a double-stage-grid connected PV generator under grid faults

N. Hamrouni

Laboratory of Analysis and Treatment of Electric Signals and Energetic Systems,
Science Faculty of Tunis-University of Tunis El Manar, University of Carthage, Tunisia

ABSTRACT: This paper presents a command of a typical double-stage grid-connected photo-voltaic system functioning under normal and disturbance operation modes. The aim of this proposed command is to contribute to system stability and ensure connection as long as possible during grid faults. Under Normal Operation Mode (NOM), the command extracts the maximum photovoltaic power and regulates, respectively, the dc-link voltage and the inverter output current. Otherwise, under grid Faulty Operation Mode (FOM), the proposed command permits to adjustment of delivered photovoltaic (PV) power with the inverter output power at the Point of Common Coupling (PCC). It changes the command signal from its Maximum Power Point (MPP) voltage to a new voltage corresponding to less power. According to the state of the grid voltage at the PCC and the inverter rating current, the control unit generates instantaneously powers that the inverter can inject to the alternative side. Those powers will be used as references to control the inverter output current, the dc-link voltage and the activation or deactivation of the MPPT. A case study of a 3kW solar generator simulated in Matlab/Simulink software is used to illustrate the proposed control.

1 INTRODUCTION

The connection of the PV generator (PVG) to the grid raises several voltage problems in the PCC. These problems are generally voltage deviations like voltage amplitude drop and higher harmonic components. According some technical literature, the main problems are the voltage dips. This disturbance has a negative impact on the stability and the power quality of the grid connected PV system. The common causes for these voltage dips are principally short circuits and starts of large load. To protect the overall system, the control of the inverter must disconnect the PV generator from the grid. In Camieletto et al. 2011, it is required that the energy source ceases to provide powers to the grid network when voltage problems occur in the PCC. The loss of the photovoltaic generator at the first time of the grid disturbances is not an optimal solution. Moreover, the repeated disconnections of the grid present a negative impact on the system stability. To maintain the connection as long as possible, in spite of grid problems, several command methods are studied (Blaabjerg et al. 2006; Rodriguez et al. 2011). Among several studies for voltage dips, the control strategy developed in Azevedo et al. 2009 using proportional-resonant current control, allows the system to operate without an over-current and overcomes the high-level harmonics when voltage dips occur at the PCC. Reactive power support was not considered in this study. A method was presented in Mirhosseini et al. 2013 to control the voltage sequences. Two controllers are implemented; one for the positive and the second for the negative. This study presented dynamic limitations of using this control configuration due to the delays introduced by controller loops. In Banu et al. 2014, a study on double-stage based three-phase grid connected PV systems under several grid faults is presented. It used an external control loop that regulates dc-link voltage and an internal control loop that regulates active and reactive current components of the grid current. In (Seo et al. 2009), the control of the ac-side was developed. The impact of grid faults on the photovoltaic current and voltage was shown in this study. In Pannell et al. 2013 and Silva et al. 2014, the grid connected PV system command was based on passive control, e.g., crowbar and chopper resistive. However in (Leon et al. 2012) the commands are based on active control schemes. Those studies showed that the active command provides to the inverter an FRT capability but the passive methods have the drawbacks of requiring additional components and dissipating significant power during the voltage sags. For a single-stage PV plant configuration, some research was done in Mirhosseini et al. 2013 evaluating the FRT issues under unbalanced grid voltages. Whereas in the application of two-stage grid connected PV system, no study has discussed a method to protect the system during grid faults. Moreover, these papers discussed dynamic performances of the overall system under grid sags, but they haven't studied the interaction between the grid fault and the MPPT control. They partially discussed only the inverter control and omitted the MPPT algorithm. In this study, the command approach developed in Sang Won et al. 2016 for a double stage grid connected PV system is adopted, although the MPPT, the active and reactive (PQ) controllers are modified based on the level of grid voltage droops. The main propose of this study is to develop a control strategy that makes the system able to withstand any type of voltage dips while injecting reactive power to meet the grid codes. According to the level of the grid voltage dip, the command approach changes the reference voltage from its MPP voltage to a new voltage in order to inject a less power into the grid and protect the dc-link capacitor. The deactivation of the MPPT command ensures the safety and the stability of the overall system in connected functioning mode. This rest of this paper is divided in two parts. Part 2 describes the control strategy of the grid connected system under grid voltage sags. Part 3 presents the principal simulation results and discussions.

2 CONTROL STRATEGY

Several grid-connected PV systems are presented in the technical literature. Due to their advantages in achieving higher energy conversion efficiency, modularity and power density, the study is focused on the two-stage grid connected PV system. The diagram of the system is shown in Figure 1. It consists of PVG, dc-dc converter, three-phase inverter and L-filter. The dc-dc converter is adopted to boost-up the PV voltage within an acceptable range of the PV inverter and makes the system flexible to track the MPP of the PVG. The inverter achieves the control of its output powers according to the level of the voltage dips. The L-filter is used to couple the inverter to the ac grid and to provide low current THD performance to comply with the grid codes.

A control unit composed of a voltage dip detection, a power bloc, a current and voltage controllers and an MPPT command is studied and proposed arround the converters. This unit permits us to control the overall system in its various states. In normal operation mode (NOM), the control unit achieves simultaneously the maximal power tracking, regulates the dc-link voltage and control the currents injected into the grid. Its objective is to deliver only the active power to the alternative

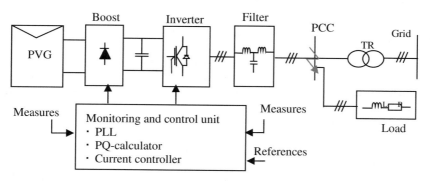

Figure 1. Hardware schematic configuration and control of a two-stage grid connected PV system.

side. The reference active power is approximately equal to the maximum PV power ($P* \approx P_{pvmax}$) while the reference reactive power ($Q*$) is equal to zero. The control system should switch from the normal operation to the grid fault operation once the amplitude of the voltage sag exceeds its limits. In this situation, the MPPT function based on Perturbation and Observation (P&O) is deactivated and the inverter should deliver reactive power according to the level of the voltage dip and the inverter rating current. To maintain the grid-connected photovoltaic system stability and avoid tripping the inverter over-current protection, the PV power should match with the active power provided by the inverter.

2.1 Boost control

In NOM, the P&O-MPPT operation is ensured by the boost converter. It adjusts the output voltage of the PVG (v_{pv}) to its MPP voltage ($v_{pv}*$). The error between the MPP voltage and the PV voltage goes through the R_v controller and generates consequently the reference capacitor current. Subsequently, by controlling the inductor current via the R_c controller, the reference inductor voltage is generated. Proportional correctors are used to control the PV current and voltage. In order to improve the performances of the PV current and voltage controllers further, the tuning of the controllers is adaptively changed based on a pre-calculated look-up table according to the real time value of the climatic conditions (Table 1).

In order to maintain the grid connected PV system functioning under grid faults, some modifications are realized. In this operation mode, the approach command matches the inverter output power to the PV power and keeps consequently the dc-link voltage constant. The instantaneous Symmetrical Grid Voltage Dip (SGVD) increases the current delivered by the inverter in order to keep the power balanced. In fact, if the active power injected into the grid (P_{inv}) decreases while the MPPT command continues to extract the maximum PV power (P_{pvmax}), the difference between production and consumption will be stored in the dc-link capacitor. This will lead to an increase in the dc-link voltage which cannot be maintained by the inverter at the desired value under grid faults. In two-stage grid connected PV systems, two ways to limit the dc-link voltage under grid faults are presented in the technical literature. The first solution short circuits the PVG by tuning ON the switch of the boost throughout the voltage dip duration and the second method leaves the PVG open by tuning OFF the switch of the boost converter. Those solutions explained the next stop of the transfer of the PV energy to the dc-link bus, however the dc-link voltage keeps itself regulated by the voltage control loop. In this paper, the connection is ensured and the developed P&O-MPPT algorithm should be deactivated in order to supply less PV power equal to the inverter output power when SGVD occurs at the PCC. The functioning power point of the grid connected system moves to a lower power level in order to protect the semiconductor switches and to avoid dc-link overvoltage. In Azevedo et al. 2009, a positive Δv_{pv} is added to the V_{pvmax} in order to make the displacement of the MPP to a new operating point with less power. The displacement Δv_{pv} is ensured by the PI regulator that regulates the dc-bus voltage to the reference value. In our study, the boost is controlled to find a new reference voltage ($v_{pv}*$) that ensured a balance power. According to the P-V characteristic given by Figure 2, two operation points (A and B) are possible. The first is at low voltage (on the left-side of MPP); however, the second is at high voltage (on the right-side of MPP). Each of these operation points presents advantages and disadvantages. In fact, the moving of the MPP to the right-side is faster than to the left side. Moreover, there is a higher oscillation of PV power at the right-side compared to the left side. On the right-side, the operation point can go beyond the open circuit voltage (V_{oc}) under fast changing of climatic conditions. Therefore, the

Table 1. Principal parameters of the grid connected PV system.

Φ (kW/M^2)	0.2	0.3	0.4	0.5	0.6	0.7	0.8	0.9	1
k_c	3	6	9	12	16	19	23	27	34
k_v	15.2	15	14.8	14.1	14	13.2	12	10.5	10

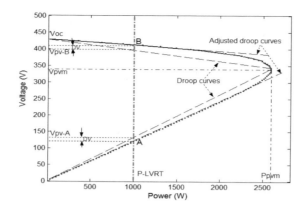

Figure 2. P-V characteristic of the PVG, the droop curves and operation points.

PVG enables to deliver power to the grid. In this case, the command should perturb the operation point in order to move it to the left-side. Otherwise, on the left-side, the MPP presents a poor control and a large oscillation of PV current and voltage when high grid sags occur at the PCC. According to the P-V characteristic, the dynamic performance is slower, whereas the oscillation of the PV power is less than at the right-side. In fact, the use of two-stage configuration allows the possibility to operate at the two-sides. This possibility is limited in the single stage because the operation point should usually move to the right-side of the MPP. To overcome the drawbacks, a Constant Power Generation based Artificial Neural Network (CPG-ANN) algorithm is developed to control the PVG when faults occur at the PCC.

According to Figure 2, the photovoltaic power is given by:

$$P_{pv} = \begin{cases} P_{pvm} \\ P_{LVRT} \end{cases} \tag{1}$$

P_{pvm} and P_{LVRT} are the PV powers provided, respectively, by the generator under normal grid condition and symmetrical grid voltage. The control strategy of the grid connected PV system is based on the assumption that the photovoltaic power is equal to the inverter output power (conduction and switching losses are assumed to be zero). Hence, the corresponding reference voltage is given by:

$$V_{pv}^* = \begin{cases} V_{pvm} \\ V_{pv} \end{cases} \tag{2}$$

V_{pvm} is the reference voltage given by the MPPT-ANN algorithm under normal conditions, however, V_{pv} is the reference voltage given by the CPG-ANN algorithm under grid faults. During the CPG mode, the MPP is necessary to calculate the operating point both at right and left sides. The control structure of the dc-dc converter is shown in Figure 3. According to Figure 2, both A and B are the operation points under grid faults. The grid connected PV system can be controlled at both points. Therefore, the PVG must reduce its power injection under grid voltage sags. As indicated by Figure 2, the P-V characteristic on the right-side of MPP can be supposed to be linear. Under grid sags, the operation point should move from the MPP (P_{pvm}, V_{pvm}) to a lower power point named B (P_{LVRT}, V_{pv-B}). According to the Side Splitter theorem and the assumption given previously, the new reference voltage is given by:

$$v_{pv-B} = \frac{P_{LVRT}}{P_{pvm}} (P_{pvm} - V_{oc}) + V_{oc} \tag{3}$$

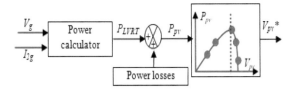

Figure 3. Diagram scheme of the CPG command.

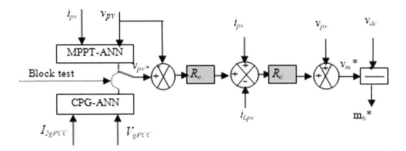

Figure 4. Control structure of the boost converter.

2.2 Inverter control

The aim of the inverter control in normal operation mode is to transfer the maximal photovoltaic power to the alternative side, to control the dc-bus voltage and to inject alternative currents with low THD under unity power factor. As indicated by Figure 4, the proposed command scheme around the three-phase inverter is composed of two loops. The first regulates the dc-bus voltage and the second controls the inverter output currents. The compensation of the reactive energy required by the local load is ensured by the electrical network. Under these conditions, three elementary blocks of command are necessary; the first is the Phase Locked Loop (PLL), the second concern the dc-link voltage regulator while the third is the inverter output current regulator.

Under grid voltage disturbances, the inverter command switches from the NOM to faulty operation mode (FOM). The reference powers (active and reactive) should be changed within the voltage sag level. In NOM, the first unit of the voltage loop supplies a dc-current which permits to determine the reference active power requested by the alternative side. The strategy is based on the assumption that the maximum PV power is equal to the inverter output power injected into the grid. It is given by $P^* = i_e * v_{dc} \approx P_{pv\,max}$ whereas the reactive power ($Q*$) should be equal to zero. The goal of the voltage loop is to ensure dc-link voltage constant and to provide a reference current (i_e*). The reference current is given by:

$$i_e^* = i_{dc} - k_{dc}(v_{dc}^* - v_{dc}) \qquad (4)$$

To regulate the dc-link voltage, a proportional–integral (PI) regulator has been used. Under grid voltage dip, reference powers (P* and Q*) should deviate. Their instantaneous values are given by:

$$\begin{cases} P^* = \frac{3}{2}V_{LVRT}I_{2d-LVRT} \approx P_{pv}^* \\ Q^* = \frac{3}{2}V_{LVRT}I_{2q-LVRT} \neq 0 \end{cases} \qquad (5)$$

The second unit of the voltage regulation provided the dq-currents ($i_{2d}*$, $i_{2q}*$) which are used as references to control the inverter output currents. These current components are calculated from

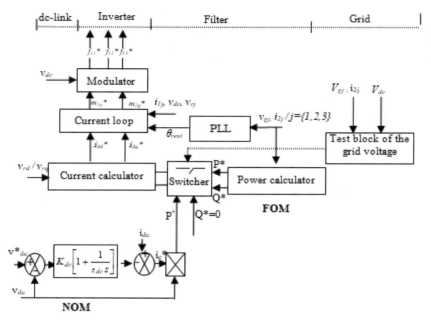

Figure 5. Hardware schematic control of the inverter.

the grid voltage and the active and reactive power references. They are expressed as follows:

$$\begin{bmatrix} i_{2d}* \\ i_{2q}* \end{bmatrix} = \frac{1}{V_q^2 + V_d^2} \begin{bmatrix} P* & -Q* \\ Q* & P* \end{bmatrix} \begin{bmatrix} V_d \\ V_q \end{bmatrix} \qquad (6)$$

The PLL loop generates the instantaneous values of the grid voltage frequency and phase at the PCC. The PLL loop should be designed to respond within a minimum of overshoot under grid disturbances. The PLL became active when the difference between grid phase angle (θ_{rest}) and inverter angle (θ_{ond}) is reduced to zero. The current controller is composed of individual regulators for active and reactive parts with additional compensation terms. The current controller should drive the inverter output current to the reference currents, namely to the shape and phase of the utility grid voltage. The current regulator is common for normal and faulty operation modes.

3 SIMULATION RESULTS

In this section, attention is devoted to the symmetrical grid voltage dip (SGVD) influence on the grid connected PV system when the MPPT is deactivated.

In order to investigate the performances of the approach control, the behavior of the system is examined under a 50%-100 ms three phase voltage dip (Cond. I) between 1 and 1,1 s. In addition to that, the dynamic performance of the proposed control approach is also examined under various solar irradiances which changes from 1000 W/m² to 400 W/m² at 1.35 s (Cond. II). The principal parameters of system and regulators are given in Table 2. The grid voltage and the inverter output current at the PCC are given by Figure 5(a). A low displacement appears during the SGVD between the inverter output current and the ac-side line voltage. As a result, the power factor leaves the unit and the inverter generates, instantaneously, a reactive power. Before t = 1s, the system is under normal operation mode; the irradiance and the ambient temperature are equal to 1000 W/m² and 25°C, respectively. In this condition, the injected average reactive power is zero and the active is

Table 2. Principal parameters of system and regulators.

Value	Symbols	Parameters
0.01 H	L	Inductance of the filter
0.92 μF	C	Capacitance of the DC-link
0.05 Ω	R	Resistance of the filter
314 rd/s	W_r	Pulsation of the grid voltage
515 V	U_{cref}	-link voltage reference
10	K_{id}	Gain of the C_{id}-Current regulator
0.001	τ_{id}	Const. time of C_{id}-Current regulator
10	K_{id}	Gain of the C_{aq}-Current regulator
0.001	τ_{id}	Const. time of C_{aq}-Current regulator
5	k_{dc}	Gain of dc voltage regulator
0.02	τ_{dc}	Const. time of dc voltage regulator
10	k_θ	Gain of PLL regulator
0.1	τ_θ	Constant time of PLL regulator

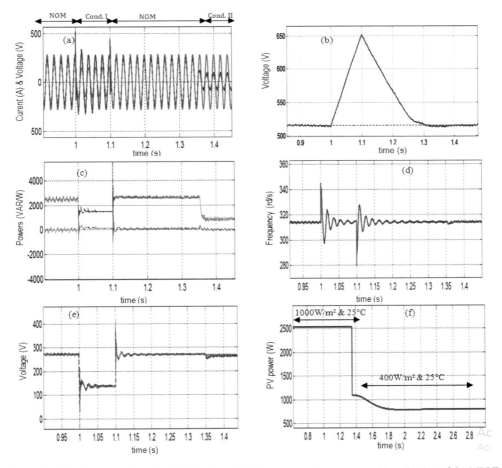

Figure 6. Simulation results under 50%-100ms SGVD and various irradiances with activation of the MPPT control: (a): single phase grid voltage and (.*40) current; (b): dc-link voltage (measure and reference); (c): Inverter output active (blue) and reactive (green) powers, (d): grid frequency provided by the PLL, (e): inverter voltage d-component (measure and reference); (f): maximum PV power.

maximal ($\approx 2560\,W$). During the voltage dip which occurs at $t = 1\,s$, the command continues to extract the maximum PV power and regulates the inverter output powers. The inverter output active and reactive powers deviate from their optimal values. The active power decreases while the reactive power increases in order to withstand the SGVD (Figure 5(c)). In this condition, the inverter output current increases until the maximum limit inverter current is reached (Figure 5a). According to Figure 5(b), the dc-link voltage rises to a high level which can trigger the dc-voltage protection. At $t = 1.1\,s$, the SGVD is suppressed and consequently the voltage decreases progressively discharging the extra energy into the ac-side. The dc-link voltage reaches its reference in 200 ms (Figure 5b). At $t = 1.35\,s$ the irradiance decreases from 1000 to $400\,W/m^2$, the amount of the injected power is reduced to 800 W and the reactive power remains equal to zero (Figure 5c). Consequently, the inverter output current is reduced (Figure 5a). As indicated by Figure 5f, the PVG provided its maximum power during the fault time because the MPPT is activated. As it can be seen by this figure, the voltage dip hasn't an influence on the power level supplied by the boost converter. This is due to the dc-link capacitor which acts as a buffer zone between 1 and 1.1 s. During the grid faults and the variation of the climatic conditions, the measure voltage d-component of the filter, as depicted in Figure 5(e), follows its reference provided by the current controller. It shows the performance of the grid voltage controller in faulty operation modes (Cond. I) and variable climatic conditions (Cond. II).

4 CONCLUSIONS

The paper presents a command strategy approach to control the overall system functioning under normal conditions and symmetrical grid voltage dips. It permits during the NOM to rise the PV power to a suitable level corresponds to the MPPT. Moreover, it regulates simultaneously the dc-link voltage and the current provided by the inverter. It permits injection into the grid a current with low harmonic distortion and active power at the unity power factor. However, under SGVD, the power references have been deviated within the voltage sag level. Consequently, the MPPT command is deactivated and the reactive power became slightly greater than zero. It changes the MPP to less power compared with pre-fault conditions. Finally, we demonstrate that the command approach allows grid connected PV system operation under grid dips and ensures grid sinusoidal current. Moreover, it provides ride-through capability during grid faults. The command improves the performance of the grid connected PV system (stability and reliability). Moreover, it overcomes some problems as well as reduces the PV power to the target value delivered by the inverter under grid faults.

ACKNOWLEDGMENT

This work is part of the project "Projets Jeunes chercheurs" funded by the Ministry of Higher Education and Scientific Research. The support of the ministry is kindly acknowledged.

REFERENCES

Azevedo M. G. S., Rodriguez P., Cavalcanti M. C., Vázquez G., Neves F. A. S., 2009, New control strategy to allow the photovoltaic systems operation under grid faults. *Power Electronics Conference, Brazilian. pp. 196–201.*

Banu I. V., Istrate M., 2014. Study of three-phase photovoltaic system under grid faults, *International Conference and Exposition on Electronics and Power Engineering (ICEEPE). Romania.*

Blaabjerg F., Teodorescu R., Liserre M., Timbus A. V., 2006. Overview of control and grid synchronization for distributed power generation systems. *IEEE Transaction on Industrial Electronics. Vol. 53, pp. 1398–1409.*

Carnieletto R., Brandao D. I., Farret F. A., Simoes M. G., 2011. Smart grid initiative: A multifunctional single-phase voltage source inverter. *IEEE Industrial Application of Magnetics. Vol. 17, pp. 27–35.*

Leon A., Mauricio J., Solsona J., 2012. Fault ride-through enhancement of DFIG-based wind generation considering unbalanced and distorted conditions. *IEEE Transaction Energy Conversion. Vol. 27; pp. 775–783.*

Mirhosseini M., Pou J., Agelidis V. G., 2013. Current improvement of a grid-connected photovoltaic system under unbalanced voltage conditions. *5th Annual International Energy Conversion Congress and Exhibition for the Asia (ECCE), Melbourne, Australia, pp. 66–72.*

Pannell G., Zahawi B., Atkinson D., Missailidis P., 2013. Evaluation of the performance of a dc-link brake chopper as a DFIG low-voltage fault-ride through device. *IEEE Transactions on Energy and Conversion. Vol. 28, pp. 535–542.*

Rodriguez P., Luna A., Munoz A. R., Coracles F., Teodorescu R., Blaabjerg F., 2011. Control of power converters in distributed generation applications under grid fault conditions. *Proceeding of the European Conference on Cognitive Ergonomics ECCE, Rostock, Germany., pp. 2649–2656.*

Sang Wong W., Y. Yang, F. Blaabjerg, and H. Wang, 2016. Benchmarking of constant power generation strategies for single-phase grid connected photovoltaic systems, *in Proc. of APEC, pp. 370–377, Mar. 2016.*

Seo H., Kim C., Yoon Y. M., Jung C., 2009. Dynamics of grid-connected photovoltaic system at fault conditions. *Transmission Distribution Conference Exposition, Asia Pacific. pp. 1–4.*

Silva B., Moreira C., Leite H., Lopes J., 2014. Control strategies for ac fault ride through in multi-terminal HVDC grids. *IEEE Transaction Power Delivery., Vol. 29. pp. 395–405.*

Innovative and Intelligent Technology-Based Services for Smart Environments – Ben Slama et al (eds)
© 2021 Taylor & Francis Group, London, ISBN 978-1-032-02030-3

A novel 120 Hz ripple power cancelation technique in single phase DC/AC inverter for PV applications

M.F. Alsolami
Department of Electrical Engineering, Taibah University, Saudi Arabia

ABSTRACT: A novel approach for the 120 Hz ripple power mitigation in a single-phase, three-level dc/ac inverter is proposed. In the traditional approach, a bulk electrolytic capacitor is used to eliminate the low frequency ripples in a dc-link capacitor, however, this solution is not due to the low frequency components and large energy is stored. A new active filtering technique without an external circuit or extra switching power devices is proposed. As a result. a small dc-link capacitor can be realized through controlling the phase angle between different inverter ports modulating reference waveforms. Additionally, the circuit takes advantage of the superior features of GaN power switching devices, as a result cooling size will be reduced, and with faster switching speed, small size passive components can be realized. Theoretical analysis and control algorithms are provided to verify the proposed low ripple mitigation technique and the control strategy of the converter.

Keywords: Photovoltaic system, Switched Capacitor Circuit, Wide Bandgap Devices.

1 INTRODUCTION

Single-phase pulse-width modulation (PWM) inverters connected to photovoltaic (PV) panels are the trend for the future grid-connected systems. PV energy is abundant but must be effectively utilized and converted into a usable form to supply the general electrical grid (Harb et al. 2013; Sun et al. 2016). However, in single-phase grid-connected inverters, the inherent low double power (120 Hz) results in undesirable low-frequency ripples in voltage and current (Wen et al. 2012; Wang et al. 2011). And in such applications, the ripple energy is stored in an electrolytic capacitor. However, in low power applications the value of dc-link capacitance can be several Farads to satisfy the voltage variation requirements (Qin et al. 2017). This approach cannot be a practical solution for many applications, and this make the second-order frequency ripple an essential factor in determining the size of the PV inverter (Xia et al. 2017).

Different methods are proposed in literature to mitigate the second order harmonics in single-phase da/ac inverters (Gu et al 2009). In general, an external circuit as shown in Figure 1 is placed in parallel with the dc-link to decouple the low frequency ripple energy to the assigned storage elements (Krein et al. 2012). However, most of the proposed effective low ripple mitigation techniques require active switching devices besides the storage elements, such as indictors or capacitors. In Li et al. (2013), two active switching devices with one filter inductor are utilized to mitigate the ripple power. In Song et al. (2007), the capacitor is assigned to process the ripple energy in the ac side and two active power switching devices are employed in the ripple mitigation process. However, the extra switching devices and the ac capacitor will make the system less efficient, more expensive with considerable weight and large size.

In this paper, a novel approach of 120 Hz current ripple mitigation in a single phase PV system is proposed. The main advantage is to keep a constant average power at dc-link and decouple the low frequency ripple power with no need for additional switching devices. Hence, a large size film capacitor is replaced by a small size multilayer ceramic capacitor. Therefore, a smaller system size

 DOI 10.1201/9781003181545-30

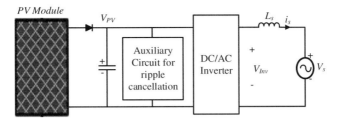

Figure 1. Traditional circuit for low ripple mitigation in single phase inverter.

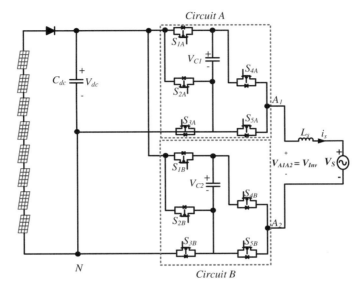

Figure 2. The proposed five-level dc/ac inverter in grid connected.

is realized, and higher efficiency is achieved with lighter weight and lower cost. Furthermore, the faster switching speed of Gan-devices contribute to further reduce the size of passive elements such as storage capacitors and output C filters (Alsolami et al. 2020). Moreover, with the high-temperature capability of Gan-devices, the size of the heatsink can be greatly reduced and with small reverse recovery time, the snubber circuit is not required.

2 THE PROPOSED THREE-PORT CONVERTER

The proposed converter in this paper will be designed for a commercial PV panel that can be implemented in residential system with a typical 120 V_{RMS} and integrated maximum power point tracking (MPPT) can have a nominal voltage of 120 Vdc.

Figure 2 shows the proposed single phase, five level dc/ac inverter. The inverter has two identical circuit structures, A and B, and each circuit consists of a boosting switching cell and half bridge inverter. The circuit A and B is composed of five switches, S1A–S5A are the assigned switching devices for circuit A, and S1B–S5B are the switching devices for circuit B. Each of the circuits A and B have only one storage capacitor, C1 and C2 respectively. The node A1 in circuit A with the node A2 in circuit B forms an ac port and the corresponding output voltage is VA1A2 (inverter voltage).

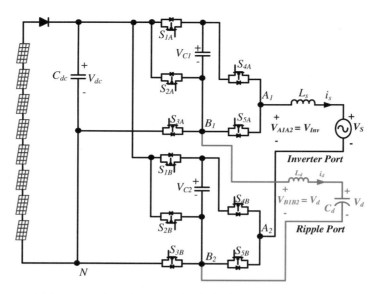

Figure 3.   Proposed dc/ac inverter in grid-connected with ripple port configuration.

Compared to other switched-mode power inverters, the proposed inverter adopts the concept of switched capacitor circuits and uses the semiconductor power devices and capacitors to build a circuit with no need for large size magnetic components. Therefore, using Gan-devices, the circuit capacitors C1 and C2 are greatly reduced, hence, multilayer ceramic capacitors will replace the large film capacitors.

A third energy ripple port with red color is configured as shown in Figure 3. The new port is assigned to process the ripple power with no need for additional active switching devices or auxiliary circuits. The important advantage of the new ripple port is the direct control over the decoupling capacitor voltage and current. Hence, the ac capacitor Cd is totally employed to store low ripple energy. Consequently, the new proposed converter only needs a small value of dc-link capacitance, Cdc, and a small value of ac decoupling capacitance, Cd, to accomplish the task. The additional attributes of the presented topology include: 1) the ability to boost the output voltage with multilevel operation, 2) the utilization of fast switching speed of Gan-devices to minimize the circuit size, and 3) the ability to discover a new port to process the ripple power with no extra components cost or external circuit.

3   MODULATION DESCRIPTION

The PWM control method is shown in Figure 4, $V_d(t)$, is the modulating reference waveform for the ripple port whereas $V_{Inv}(t)$ is the modulating reference waveform for the inverter port. The figure shows that the $V_d(t)$ is shifted with respect to the $V_{Inv}(t)$ by phase angle ϕ. The inverter reference waveform and the ripple port reference waveforms are defined as the following:

$$\begin{cases} V_{Inv}(t) = 2m_i \sin(\omega t) \\ V_d(t) = m_d \sin(\omega t + \phi) + 1 \end{cases} \tag{1}$$

Where m_i and m_d are modulation index for inverter/ripple port, ϕ is phase difference between the $V_{Inv}(t)$ and $V_d(t)$.

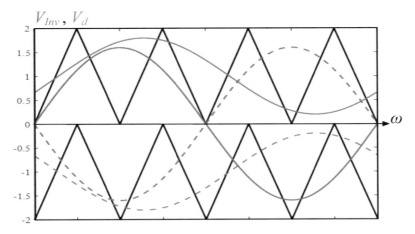

Figure 4. Carrier based PWM switches signal generation.

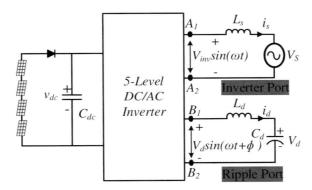

Figure 5. The proposed three port system-level diagram.

4   STRATEGY FOR 120-HZ RIPPLE POWER CANCELLATION

The system level diagram of the proposed three-port converter is shown in Figure 5. It consists of three ports: the inverter port (V_{Inv}), where the grid is connected, the decoupling ripple port (V_d), where the decoupling capacitance C_d is connected, and the dc port (V_{dc}), where the PV output voltage is connected. The inverter port voltage V_{Inv} is synchronized with the grid voltage V_s, therefore maintaining the essential synchronization properties of PV grid connection. From the diagram, decoupling ripple voltage $V_d(\omega t + \phi)$ is shifted with respect to inverter voltage $V_{Inv}(\omega t)$ with phase angle ϕ, and this angle is essential for low ripple power cancellations at dc-link. In the following section, the phase angle ϕ can be figured out through the analysis.

Equations to describe the voltage and current flow at each port in Figure 5 are derived as shown in (1), where Vs, Is, Vd and Id are the RMS amplitude of vs (t), is(t), vd(t) and id(t) respectively, δ is the phase angle difference between vs (t) and is(t), and ϕ is the phase angle difference between vd(t) and vs(t)

$$\begin{cases} v_S\,(t) = \sqrt{2}V_S \sin{(\omega t)} \\ i_S\,(t) = \sqrt{2}I_S \sin{(\omega t + \theta)} \\ v_d\,(t) = \sqrt{2}V_d \sin{(\omega t + \phi)} \\ i_d\,(t) = \sqrt{2}V_d \omega C_d \cos{(\omega t + \phi)} \end{cases} \tag{2}$$

213

The voltages across inductors Ls and Ld are very small when compared to the grid voltage, hence, they can be neglected. Therefore, when calculating the instantons power, this assumption can be: VA1A2 $\approx$ Vs, and VB1B2 $\approx$ Vd, and the instantons power $p_{ac}(t)$ can be derived as the following,

$$p_{ac}(t) = v_S(t) i_S(t) + v_d(t) i_d(t)$$

$$= -V_S I_S \cos(\theta) - V_S I_S \cos(2\omega t + \theta) + V_d^2 \omega C_d \sin(2\omega t + 2\phi) \tag{3}$$

From (3), the ac power $p_{ac}(t)$ is consisting of two components: the constant power Po and the oscillating ripple power $p_{rip}(t)$,

$$P_o = V_S I_S \cos(\theta) \tag{4}$$

$$p_{rip}(t) = V_d^2 \omega C_d \sin(2\omega t + 2\phi) + -V_S I_S \cos(2\omega t + \theta) \tag{5}$$

from the power balance theory, the input/output power are equal, hence,

$$p_{dc}(t) = p_{ac}(t) = v_{dc}(t) i_{dc}(t) \tag{6}$$

Then, based on (2) and (5), the dc current can be obtained as:

$$i_{dc}(t) = \frac{1}{V_{dc}} p_{ac}(t)$$

$$i_{dc}(t) = \frac{1}{V_{dc}} V_S I_S \cos(\theta) + \frac{1}{V_{dc}} \left(V_d^2 \omega C_d \sin(2\omega t + 2\phi) - V_S I_S \cos(2\omega t + \theta) \right) \tag{7}$$

Therefore, if the instantons current $i_{dc}(t)$ needs to be zero, the oscillating ripple part must be zero, $p_{rip}(t) = 0$, Hence, equation (5) becomes

$$V_d^2 \omega C_d \sin(2\omega t + 2\phi) = V_S I_S \cos(2\omega t - \theta) \tag{8}$$

Solving equation (8), the following equation can be derived,

$$\begin{cases} V_d = \sqrt{\frac{V_S I_S}{\omega C_d}} \\ \phi = \frac{\theta}{2} + \frac{\pi}{4} \end{cases} \tag{9}$$

Therefore, if the voltage across the capacitor V_d is controlled such that the expressions in (9) are satisfied, the dc current variation is nearly zero. Therefore, the power decoupling in the proposed inverter is realized. From equation (9), the decoupling capacitance can be calculated from the following relation,

$$C_d = \frac{V_S I_S}{\omega V_d^2} \tag{10}$$

5   SIMULATION RESULTS

A single-phase dc/ac inverter is simulated at different load conditions with a closed loop control manner. The utility grid voltage is set at 120 V_{rms}, inverter and ripple port modulation indices' values are set at $m_i = m_{d} = 0.67$, respectively. The decoupling capacitance C_d and inductance L_d are selected to be 150 μ F, 200 μ H, respectively. System frequencies are 60 Hz and DC-link voltage is 125 V_{dc}.

Figure 6 shows the waveforms of grid voltage V_S, the grid current I_S, the decoupling capacitor voltage V_d, dc side power P_{dc} and dc-link current I_{dc}, respectively. Prior to the power decoupling control activation, the current i_{dc} and the power P_{dc} is oscillating at 120 Hz due to the nature of the

214

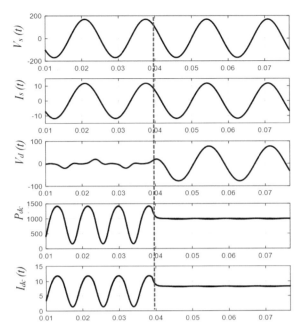

Figure 6. Simulation results before/after ripple mitigation technique is activated.

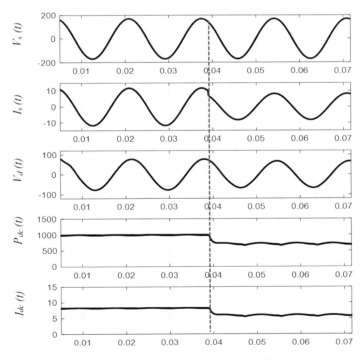

Figure 7. A step change with 30% drop in output power ($\theta = 0$, $\phi = \pi/4$).

60 Hz single-phase system. However, when the power decoupling control is activated at a time of 0.04 s, the magnitude of the second-order harmonic of the dc current (I_{dc}) is reduced to very small and the dc source current i_{dc} is approximately constant. With this condition, the angle ϕ between the two reference voltage waveforms according to equation (9), is nearly $45°$. Furthermore, the power decoupling control has no effect on grid current i_s and the power factor maintains unity as specified.

Figure 7 shows the simulation waveforms when the load has a step change. When the output power P_o has dropped to 70% of its original power demand (1 kW to 700 W). The control will track the change in the output power, hence, Voltage in the decoupling port V_d is controlled to satisfy the expressions in (8). And the voltage in the dc-link is kept constant by the outer loop controller, and the inner loop controller track the change in grid current I_s.

6 CONCLUSION

A single-phase dc/ac inverter with a low ripple power elimination technique for PV applications is proposed. The operation principle of the inverter was discussed and a novel approach for ripple power mitigation were analyzed and theoretically proved. The effectiveness of new mitigation approach was proved in simulation and the 120 Hz ripple was reduced to a very small value without adding more switching devices. This approach enables the circuit to employ small multilayer ceramic capacitors at dc-link instead of large film capacitors. Furthermore, with the fast switching speed of Gan devices, the circuit was capable of operating at a high switching frequency, thus smaller size heatsinks and passive components were achieved.

REFERENCES

M. F. Alsolami, K. Potty and J. G. Wang, "Mitigation of Double Line-Frequency Current Ripple in Switched Capacitor based UPS system," in *IEEE Transactions on Power Electronics*, 2020, Early Access Article.

L. Gu, X. Ruan, M. Xu, and K. Yao, "Means of eliminating electrolytic capacitor in ACIDC power supplies for LED lightings," IEEE Trans. Power Electron., vol. 24, no. 5, pp. 1399–1408, May 2009

S. Harb, M. Mirjafari and R. S. Balog, "Ripple-Port Module-Integrated Inverter for Grid-Connected PV Applications," in IEEE Transactions on Industry Applications, vol. 49, no. 6, pp. 2692–2698, Nov.-Dec. 2013,

P. T. Krein, R. S. Balog and M. Mirjafari, "Minimum Energy and Capacitance Requirements for Single-Phase Inverters and Rectifiers Using a Ripple Port," in IEEE Transactions on Power Electronics, vol. 27, no. 11, pp. 4690–4698, Nov. 2012

Li, H., Zhang, K., Zhao, H., Fan, S., Xiong, J.: 'Active power decoupling for high-power single-phase WM rectifiers', IEEE Trans. Power Electron., 2013, 28, (3), pp. 1308–1319

S. Qin, Y. Lei, C. Barth, W. C. Liu, and R. C. N. Pilawa-Podgurski, "A high power density series-stacked energy buffer for power pulsation decoupling in single-phase converters," IEEE Trans. Power Electron., vol. 32, no. 6, pp. 4905–4924, Jun. 2017.

Y. J. Song, S. B. Han, X. Li, S. I. Park, H. G. Jeong, and B. M. Jung, "A power control scheme to improve the performance of a fuel cell hybrid power source for residential application," in Proc. IEEE Power Electron. Spec. Conf., 2007, pp. 1261–1266.

Y. Sun, Y. Liu, M. Su, W. Xiong and J. Yang, "Review of Active Power Decoupling Topologies in Single-Phase Systems," in IEEE Transactions on Power Electronics, vol. 31, no. 7, pp. 4778–4794, July 2016,

Y. Xia, J. Roy, and R. Ayyanar, "A capacitance-minimized, doubly grounded transformer less photovoltaic inverter with inherent active power decoupling," IEEE Trans. Power Electron., vol. 32, no. 7, pp. 5188–5201, Jul. 2017.

R. Wang, F. Wang, D. Boroyevich, R. Burgos, R. Lai, P. Ning, and K. Rajashekara, "A high power density single-phase PWM rectifier withactive ripple energy storage," IEEE Trans. Power Electron., vol. 26, no. 5, pp. 1430–1443, May 2011.

H. Wen, W. Xiao, X. Wen, and A. Peter, "Analysis and evaluation of DC link capacitors for high power density electric vehicle drive systems," IEEE Trans. Veh. Teclmol., no. 99, p. 1, 2012, early Access.

*Innovative and Intelligent Technology-Based Services for Smart
Environments – Ben Slama et al (eds)*
© 2021 Taylor & Francis Group, London, ISBN 978-1-032-02030-3

Smart grid irrigation

K. Laabidi
*Department of Computer and Network Engineering, College of Computer Science and Engineering,
University of Jeddah, Jeddah, Saudi Arabia*
University Tunis Elmanar, Tunisia

M. Khayyat
*Department of Information System and Technology, College of Computer Science and Engineering,
University of Jeddah, Jeddah, Saudi Arabia*

T. Almohamadi
*Department of Computer and Network Engineering, College of Computer Science and Engineering,
University of Jeddah, Jeddah, Saudi Arabia*

ABSTRACT: The natural environment is essential to human life since it impacts people's health
and quality of life. Plants are part of this environment, and they must be preserved and protected
by irrigating them as needed. Technological solutions need to be employed, and smart irrigation
systems via the Internet of Things (IoT) can contribute to providing a healthier life in innovative
and intelligent ways. Thus, this research aims at automating irrigation for plants to create a healthy
environment for citizens. This is done by using the smart grid irrigation systems concept, which
helps farmers plant as much of the farmland as possible with less effort to obtain many of the
diverse crops that people need like food or to increase the aesthetic appeal of an area. We have
utilized IoT technologies, so we have designed and integrated an inexpensive smart grid irrigation
system using the Arduino control panel for a better quality of life.

Keywords: Environment, Smart irrigation systems, Smart Grid, IoT, Arduino.

1   INTRODUCTION

There are many important factors in agriculture, like temperature. Thus, climate change and agriculture are interlinked processes, both of which occur on a global scale and in which climate change
affects agriculture in several ways, including changes in temperature rates, precipitation, extreme
climate fluctuations such as heat waves, changes in pest populations, crop disease events, changes
in carbon dioxide in the atmosphere and ozone layer concentrations near the Earth's surface, all of
which can result in changes in the nutritional quality of some foods (Raza et al. 2019). One of the
worst phenomena caused by climate change and lack of precipitation is drought. It can have a major
impact on both the ecosystem and agriculture in affected areas. Although droughts may last for
many years, a short period of severe droughts will cause enormous damage (Delk 2020), and cause
losses to the local economy (Haque et al. 2017). This global phenomenon has a wide impact on the
field of agriculture. The impact of droughts on agriculture was estimated using the US Drought
Monitor where it was found that negative and statistically significant effects of drought on crop
yields were equal to reductions in the range of 0.1% to 1.2% for corn and soybean yields for each
additional week of drought in dryland counties and 0.1% to 0.5% in irrigated counties (Kuwayama
et al. 2019). Herein lies our need to guide water consumption. Some smart farming systems are
designed to address the water consumption problem.

The advantages of these systems vary, some work with timers and some are controlled by the
farmer; these systems have many benefits as shown in Figure 1.

DOI 10.1201/9781003181545-31

Considering these the proposed irrigation systems, which were developed to help reduce these problems, an integrated auxiliary system for plant irrigation based on its needs was developed rather than direct and random water wastage. An intelligent garden control system works with Arduino where the system outputs depend on inputs from the soil moisture sensor which senses soil temperature and humidity to test the plants' need for water and thus sends a signal to the water pump to turn it on or off, in addition to other features, such as remote pump operation control and others. Arduino Uno, the soil moisture sensor and the water pump were used in this research. See Figure 2.

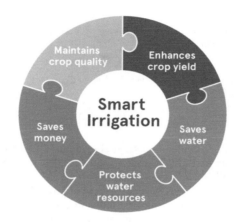

Figure 1.   Benefits of the smart irrigation systems

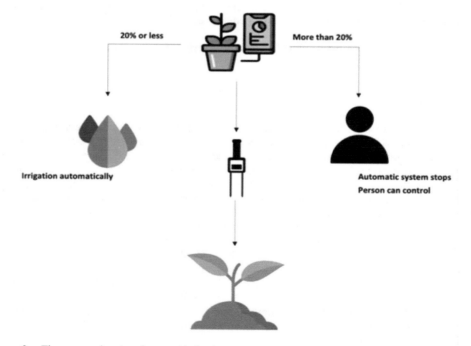

Figure 2.   The proposed system for smart irrigation.

2 SIMILAR SYSTEMS

There are some systems similar to the system in this paper, but they operate differently, such as:

1. The irrigation system with an automatic rain gun is a system that works on irrigation only in case of a severe need of water. The Android SDK program that provides the tools and software applications needed to develop the applications on the Android system was developed using the Java language, and this application uses the GPRS feature to control irrigation as it helps in reducing the agricultural cost economically (Singh & Kishore 2020).
2. A system that works on remote monitoring and control and depends on GSM/Bluetooth. In the event that the system is not monitored and controlled, it works automatically according to the temperature and humidity sensors. The system gives users some important information such as electricity conditions, temperature and water level in the soil through the GCM network (Monica 2017).
3. The IoT alarm system is based on the SIM900A module and is dedicated to global warming, as the system collects information such as temperature and humidity, and when normal levels are exceeded and imbalance occurs, alarm notifications are sent to the user's phone (Rawal 2017).

What distinguishes the system in this paper is its ability to cover large areas and diverse land types and give individual results for each specific land plot, making it easier for farmers with diverse land areas and/or crops to process irrigation with greater accuracy. It can also be used in small home gardens and by those interested in agriculture on a lesser scale.

3 METHODOLOGY

This irrigation system is based on the Arduino plate technology, which is an electronic development board consisting of an open-source electronic circuit with a computer programmable microcontroller designed to facilitate the use of interactive electronics in interdisciplinary projects. Arduino is mainly used in the design of interactive electronics projects or projects that aim to build different environmental sensors. Its main function is to measure soil moisture by means of a soil moisture sensor consisting of copper sheets as electrodes.

The soil moisture sensor has two output pins for digital and analog output. Outputs are compared with a reference value using the LM393 comparison. The digital output electrode gives a 0 low digital output when the soil is wet. Therefore, we will use the analog output from the sensor module by connecting it to one of the analog electrodes in the Arduino board, and when we use the analog output the humidity detection value can be determined and adjusted by the program (Borah 2020). We also used a solenoid valve that works on the principle of the electromagnet; When the soil moisture level reaches 20% or less, an electric current flows, and the solenoid valve opens and water starts pumping and when the soil moisture level reaches 100%, the electric current stops and the solenoid valve turns off, so the water stops pumping. We used some integral components and basic parts with the Arduino board to design the proposed model, as shown in Figure 3.

We summarized the steps for the work with an illustration in Figure 4.

4 IMPLEMENTATION AND SIMULATION

After determining the design of the proposed model and its components, we conducted simulation of the project idea to avoid errors that may occur in the implementation phase and to ensure the correctness of the method. The simulation is depicted in Figure 5.

We connected all the necessary components which are:

1. USB cable to connect the Arduino UNO board with the computer so that it is programmed and as an electrical current to operate the board.

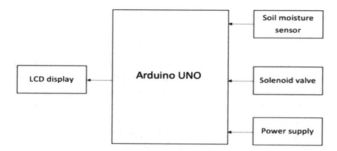

Figure 3.   Block diagram.

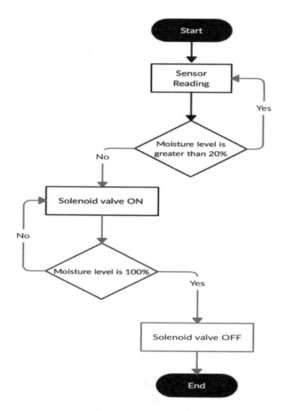

Figure 4.   Flowchart of the work.

2. Breadboard as a base for connecting electronic components to build electronic circuits.
3. Soil moisture sensor to measure soil moisture.
4. Solenoid valve to turn the water pump on or off.
5. Power supply to give power to the solenoid valve.
6. Transistor NPN as a key to control the solenoid valve.
7. Wires to connect the circuit components together and with the Arduino board.

   After we interconnected the circuit, we added the necessary programming code to operate the circuit. The result of the experiment was successful. We used the LED to test that the light turned on when the soil was dry, which meant that the solenoid valve will allow water to pass through it;

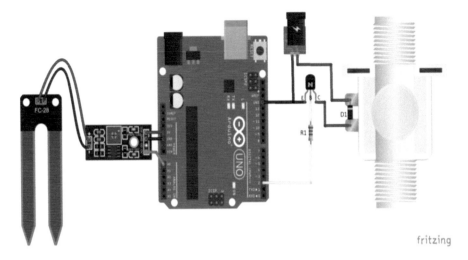

Figure 5.   Simulation of the work.

Dry
soil

Moist
soil

Figure 6.   The results.

also the light turned off when the soil was wet and which signifies that the solenoid valve will not allow water to pass through it as shown in the Figure 6.

When taking the sensor measurements and observing the interaction of the LED with the level of soil moisture, we concluded that the solenoid voltage is zero i.e. it is closed. When the soil moisture level is greater than 20%, its voltage has become 16 i.e. it is open, when the soil moisture level has reached 20% it has not closed except when the soil moisture level has increased to 100% as shown in Table 1.

Controlling the relationship between irrigation and soil drought reduces the amount of water used by relating it to the level of moisture needed in the soil for a specific crop.

This is the greatest and most important benefit of a smart grid irrigation system, since based on EPA statistics, 70% of water use occurs outside for irrigation purposes and 50% of it is wasted.

Smart grid irrigation controllers significantly reduce waste and costs and have a potential positive impact on the environment. This irrigation system also improves the health and beauty of nature since over-watering may damage it (Timberline Landscaping 2019).

221

Table 1. Test measurements.

Soil moisture (%)	Solenoid valve (V)
100	0
80	0
60	0
40	0
20	16
40	16
60	16
80	16
100	0
80	0

5 CONCLUSION

It has been found that smart grid irrigation controllers significantly reduce waste and costs, and have a potential positive impact on the environment. The irrigation system also improves the health and beauty of nature since it does not damage it by over-watering it. Thus, continuous development is important and it is possible to expand on this work by adding some features such as displaying the humidity level on the LCD display and adding stop and start buttons for the automatic mode, allowing the user to stop the automatic mode when it is not needed or when preferring manual irrigation for a given circumstance. Furthermore, a remote-control feature could be added to make the system more flexible.

REFERENCES

Borah, S.,Kumar, R., Pakhira, W., and Mukherjee, S. "Design and Analysis of Power Efficient IoT Based Capacitive Sensor System to Measure Soil Moisture", *International Conference on Computational Performance Evaluation (ComPE)*, 2020: IEEE, pp. 1–7, 2020.

Delk, J. "A Statistical Analysis of Drought and Its Global Impact," 2020.

Kuwayama, Y. Thompson, A., Bernknopf, R., Zaitchik, B., and Vail, P. "Estimating the impact of drought on agriculture using the US Drought Monitor," *American Journal of Agricultural Economics"*, vol. 101, no. 1, pp. 193–210, 2019.

Mahmudul, M. Haque, A. Amir Ahmed, and A. Rahman, "Drought Losses to Local Economy, Ch. 33 in Handbook of Drought and Water Scarcity, Vol. 1: Principles of Drought and Water Scarcity, Ed. By Eslamian S. and Eslamian F., Francis and Taylor," ed: CRC Press, USA, 2017.

Monica, M., Yeshika, B., Abhishek, G., Sanjay, H., and Dasiga, S. "Iot based control and automation of smart irrigation system: an automated irrigation system using sensors, GSM, Bluetooth and cloud technology", *International Conference on Recent Innovations in Signal processing and Embedded Systems (RISE)*, 2017: IEEE, pp. 601–607, 2017.

Rawal, S. "IOT based smart irrigation system", International Journal of Computer Applications, vol. 159, no. 8, pp. 7–11, 2017.

Raza, A. et al., "Impact of climate change on crops adaptation and strategies to tackle its outcome: A review," Plants, vol. 8, no. 2, p. 34, 2019.

Singh, D and Kishore, R. "Geometrical model for determining soil water content under sprinkler and raingun irrigation system", *International Journal of Agricultural Engineering*, vol. 13, no. 1, pp. 36–41, 2020.

Timberline Landscaping "5 Benefits of smart irrigation controller", 2019, Online Available at: https://www.timberlinelandscaping.com/5-benefits-of-smart-irrigation-controllers/.

Innovative and Intelligent Technology-Based Services for Smart
Environments – Ben Slama et al (eds)
© 2021 Taylor & Francis Group, London, ISBN 978-1-032-02030-3

Smart firefighting system for better quality of life

K. Laabidi
*Department of Computer and Network Engineering, College of Computer Science and Engineering,
University of Jeddah, Jeddah, Saudi Arabia
University Tunis Elmanar, Tunisia*

M. Khayyat
*Department of Information System and Technology, College of Computer Science and Engineering,
University of Jeddah, Jeddah, Saudi Arabia*

K. Alatoudi
*Department of Computer and Network Engineering, College of Computer Science and Engineering,
University of Jeddah, Jeddah, Saudi Arabia*

ABSTRACT: Smart buildings/cities are among the most innovative solutions for engineers to
ensure social and environmental responsibility and provide safe and secure environments for occupants. These solutions are provided by the security materials and technological devices used in these
smart buildings/cities. Smart buildings and smart cities can result in a distinctive city façade which
attracts tourists and real estate investors, thus giving the city economic advantages. For the success
of projects of this kind contribute to preserving the spirit of humanity, we propose an intelligent
system which connects between building management system (BMS) and the Internet of Things
in the building's internal structure, as the integration of these two components results in superior,
safer buildings.

Keywords: Intelligent buildings, Smart Grid, Arduino, IoT, BMS.

1   INTRODUCTION

Cities around the world are racing to provide their citizens with smart and efficient cities through
electronic, financial, agricultural, security, and other means. This requires attention and input from
multi-stakeholder groups. After careful review of statistics provided by the relevant authorities on
fires and fire extinguishment, became clear that from the year 1439 to 1441 that extinguishing
operations increased on average (high); thus research was engaged in to understand the reasons
for this rise, and to determine ways to lower the rate of building fires through the implementation
of smart buildings/cities. The relevant authorities have a clear vision of the event, which enables
then to plan innovative and enhanced ways to reduce the risks of viewing the fire by enabling the
possibility of viewing the status of the fire (event) to be remotely viewed with a three-dimensional
feature. Among the problems that were noted that led to the occurrence of these environmental
disasters, and the subsequent economic problems and loss of human lives, was the lack of innovative
ways to solve the problem of dealing with fires (for the relevant authorities to extinguish the fire).
Some of the problems that may hinder them were found to be as follows:

o relying on reports from people close to the event; this problem can be solved by modern smart
   methods.
o The lack of sufficient information on the fire, which if available may shorten their response
   time.
o Lack of complete picture of the fire location and situation, which makes it difficult for responders
   to know how much equipment and how much manpower they need.

In addition, there have been some fire events that have affected the world on a global scale on an environmental level and maybe these studies can find ways to prevent them or minimize the subsequent damage:

- Amazon forest fires in 2019 (Staff 2020)
- Australia's forest fires in 2019 (BBCNews 2020)
- Al-Haramain Train Fire in Saudi Arabia 2019 (Alarabiya 2019)

Most of these are fires that caused environmental and material damage in the areas in which they occurred. There must be solutions that are smart and flexible that serve to reduce these risks from fire and preserve human life. Therefore, this research seeks to clarify the challenges and how they can be alleviated through a review of developed studies and others. This paper is organized as follows; Section 2 provides a definition of smart buildings, gives the background of the "smart fire alarm" project under consideration discusses how these buildings are constructed from the engineering perspective, and outlines the difference between them and ordinary buildings. Section 3 describes some common types of traditional and smart fire extinguishing systems. Section 4 outlines the project's implementation through the electrical circuit, and the experiment's results are presented. Section 5 programs the primitive form of the device that we are researching.

2 LITERATURE REVIEW

2.1 Smart building definition

The term smart building does not have a specific, explicit definition or meaning. A smart building can be defined as: a structure which uses automated processes to control building operations, including ventilation, lighting, security, safety, and other systems. Thus, it encompasses the merging of two different components: technology and a built structure.

2.1.1 Types of smart buildings

There are many different types of smart buildings, and each type has its own characteristics and, to some degree, most of the considerations that have been mentioned can be applied. The main part of each smart building is the system control, which functions to control the mechanical and electrical components of the structure such as ventilation, energy systems, restarting systems, and safety systems and others (Rameshwar et al. 2020). In reverse conventional buildings subsystems work separately, and communication between them is limited, see Figure 1 (Bajer 2018), so data is within the boundaries or subsystem. It cannot be fully utilized, and it may not be easy to manage or maintain it. This shortage may be the reason for the lack of integration.

2.2 Interactive systems for smart building

Smart buildings may drive a fundamental change through the facilitation of maintenance, increased user satisfaction and well-being, –and reduction of the environmental impact of the building through sustainable architecture). Smart buildings use information technology to link a variety of subsystems in the way that these subsystems can share information to improve the performance of the building (see Figure 2). There is a misconception that smart buildings are limited only to the building automation system (BAS); however, they are more than a set of automation systems. Smart buildings can also use the Internet of Things (IoT) concept and use complex algorithms or artificial intelligence to provide functions or services to users. With advanced control and diagnostics, smart buildings enhance the occupants experience with their services in terms of security, protection, physical security, sanitation, etc. So, the use of smart buildings particularly in sectors such as business, hospitals, educational facilities, etc. has received a great deal of interest and research focus that shows that the smart building technology market will achieve global revenues of $ 8.5 billion in the year 2020, up from $ 4.7 billion in 2016, and showing that has been growing at an average annual rate of 15.9% during the forecast period (Rameshwar et al. 2020; Tracy 2016).

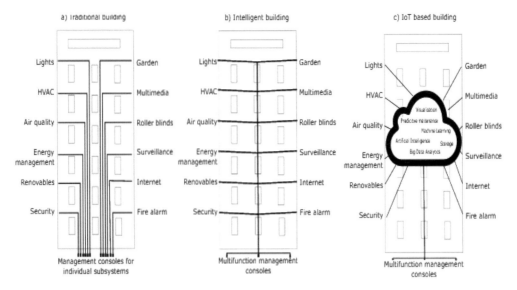

Figure 1. Regular vs. Intelligent building (Bajer 2018).

Figure 2. Smart building data integration layers (Bajer 2018).

3 METHODOLOGY

The method used to implement this device "smart fire alarm" is as follows: In the first stage of implementation, research was conducted to determine the materials needed to implement the network; Arduino UNO, Flams Sensor, 2 LED, Buzzer, Breadboard, Jumpers, NEO-6MGPS model and Ariel. These components assist in the implementation of the device through the Adriano system for fire alarm, identification of flames and of the location of the fire. When running (Flams Sensor), we need to connect (5V & GND) to the VCC port. In the second step, the Baines stage, we will deliver the d0 to it via the p0 port. Step Three (buzzer) to the buzzer will sound when notification of flames is received. See Figure 3.

The required steps to realize the software part are as follows:

1. creation Valued that is responsible for the implementation
2. calling them from the Arduino library:
   pinModeCALLING like (buzzer, OUTPUT)
   pinModeCALLING Red, OUTPUT)

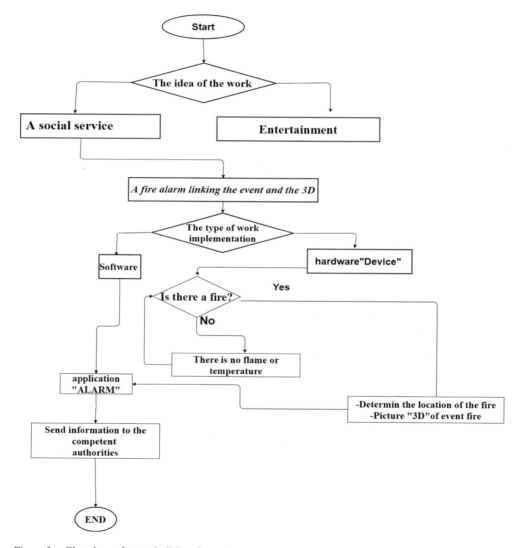

Figure 3.   Flowchart of smart building fire system.

pinModeCALLING Green, OUTPUT)
pinModeCALLING Flame, INPUT)
Serial.begin(9600)
3. implementation inside an organized matrix such as:

4 RESULTS AND DISCUSSION

Throughout the ages roads and building equipment has undergone many changes; in the time of the Roman Empire, fire extinguishing required many guards responsible for putting out fires, which were extinguished by passing a bucket from hand to hand to deliver water to the fire. Fire alarm systems appeared with the invention of the telegraph and today modern warning systems are ubiquitous and serve as warning systems (Rameshwar et al. 2020). in innovative and powerful ways. With the coming of the era of technology utilizing smoke detection and automatic extinguishing by sensors has become available, so smart buildings with these types of technology are in great demand for those concerned with security and safety. Our work is distinguished from other papers in the field of technology. First, we proposed an intelligent fire alarm device; this device senses the fire from the flame resulting from the fire, and measures the temperature of this flame. This is accomplished through the sensors available in the smart building. These sensors facilitate identification of the location of the fire through GPS and are distinguished by their ability to describe the state of the building from all sides by way of the (3D) feature, including the duration of the fire and the source of the fire in order to save time and effort and to increase the speed on the part of the relevant authorities to extinguish the fire. It also has the potential to mitigate false reports that may take waste time and endanger human life by wasting resources that may be needed elsewhere (see Figure 4). After testing and actually implementing the proposed smart grid, we reached these results: The flame sensor can sense fire and sunlight by wavelength in the range 1100–760 nm. It can quickly detect flames—fire detection within 60°C.

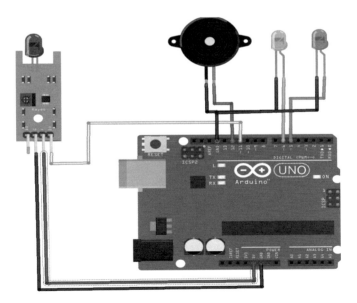

Figure 4.   Implementation under software environment.

5 CONCLUSION

During this paper, we have discussed the potential of smart buildings to enhance occupant experience and increase safety and sustainability. We have also proposed the need to continuously develop the "Internet of Things" and exploit its innovative and influencing nature in the world of technology, especially with technology engineers, who influence change for societies or developed countries in all economic and tourism aspects. We create a smart and safe environment with advanced technical methods using the Internet of things. Traditional buildings that depend on the BMS are in a circle of innovation with a strong system that needs extension. IoT devices can be a natural extension of these systems. Another distinguishing feature of the Internet of Things is integration with the cloud. This would likely allow adding more complex logic to the system, performing predictive maintenance or remote access. It is important to note that continuous monitoring and control of multiple buildings and assets would support the detection of operational inefficiencies and potential system failures, reducing maintenance and operating costs. IoT technology can be used to augment existing building management technology rather than replace it, and it can be used to consolidate building blocks at low complexity and cost (Kozłowski et al. 2017). Research findings show that technology has a great potential to help build an improved future.

REFERENCES

Alarabiya, Saudi Civil Defense fully controls fire at Haramain high-speed rail station. 2019; Available from: https://english.alarabiya.net/en/News/gulf/2019/09/29/Fire-breaks-out-at-train-station-in-Jeddah.

Bajer, M. IoT for smart buildings-long awaited revolution or lean evolution. in 2018 IEEE 6th International Conference on Future Internet of Things and Cloud (FiCloud). 2018. IEEE.

BBCNews. Australia fires: A visual guide to the bushfire crisis. 2020; Available from: https://www.bbc.com/news/world-australia-50951043.

Kozłowski, S., et al., Project Solaris, a Global Network of Autonomous Observatories: Design, Commissioning, and First Science Results. Publications of the Astronomical Society of the Pacific, 2017. 129 (980): p. 105001.

Rameshwar, R., et al., Green and smart buildings: A key to sustainable global solutions, in Green Building Management and Smart Automation. 2020, IGI Global. p. 146–163.

Staff, R. Fires in Amazon forest rose 30% in 2019. 2020; Available from: https://www.reuters.com/article/us-brazil-amazon-fires-idUSKBN1Z804V.

Tracy, P. What is a smart building and how can it benefit you? 2016 [cited 2020; Available from: https://www.rcrwireless.com/20160725/business/smart-building-tag31-tag99.

Innovative and Intelligent Technology-Based Services for Smart Environments – Ben Slama et al (eds)
© 2021 Taylor & Francis Group, London, ISBN 978-1-032-02030-3

Implementation of a cryptography algorithm for secured image transmission

A. Samoud
Systems and Signal Processing Laboratory, Science Faculty of Tunis, Tunisia

A. Cherif
FST, University of Tunis Manar, Tunisia

ABSTRACT: In this work, we will present an RSA algorithm designed for encryption and remote transmission of medical images with data security transfer. This interface is implemented using the MATLAB environment. The implementation of the image cryptography system uses the RSA algorithm with a 64 bit private key length. The transfer of images in a secure way for medical diagnosis is insured by generating a watermarked key to encrypt the original image. Besides that, we introduced a comparison study between the obtained performances and those computed with other algorithms such as AES, DES and IDEA.

1 INTRODUCTION

Nowadays, data can be misused in serious ways (access to credit cards, the transactions in e-commerce, espionage of the secret information in the military domain) especially through transmissions on the internet. It is necessary to find a robust method to secure the data during transmission (Borie & Al 2002). The encoding (ciphering) can bring solutions to this problem much more than watermarking due to difficulty of breaking the encoding key. The objective of this work is interested in the study of an asymmetric encoding methods applied for medical images by using the RSA algorithm. The objective is also to decipher it even in the presence of various types of attacks (Puech 2004).

Indeed, the university hospital centers uses and exchanges several sizes and formats of images relative to patients whose contents can possess confidential information, whether it is for diagnosis or at the personal level. Especially since these images can be remotely exchanged if the centers are interconnected by a LAN.

2 METHODS AND MATERIAL

The most common techniques are encoding of Vigenère in a single key, the symmetric algorithm DES with secret keys and the algorithm RSA with public and private keys. The algorithm RSA, at present, is the most successful technique for ciphering keys and passwords or numbers. The key varies from 64 to 1024 bits (Rivest 1978). It has the advantage of being difficult to break if the key is rather long and on the other hand, it does not need to pass on the key deprived via the network to the receiver. The principle of encoding is based on an acquisition of the image followed by a compression then a segmentation in blocks of L pixels (in normal mode L = 8 pixels or 64 bits).

Then every block of quantified data is coded by the public key (e, n) of the transmitter according to the algorithm in Figure 1. At the level of the receiver, the deciphering is made by the private key (d, n), (Chang & Al 2001).

DOI 10.1201/9781003181545-33

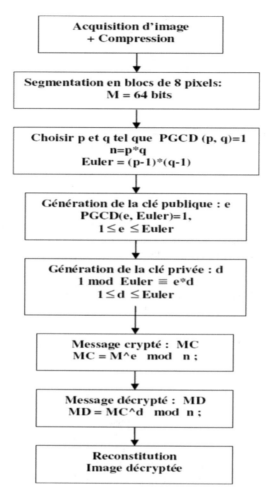

Figure 1. RSA algorithm.

3 RESULTS AND DISCUSSION

We developed using MATLAB an interface in which we implemented the algorithm RSA as well as the module of treatment and compression of the image. The image is compressed first of all according to the JPEG standard by DCT. The encoding is made in the field of transformed blocks that are 8×8. Only the coefficient BF will be taken into account to relieve the time of encoding-deciphering and decrease the image size (Obo et al. 1997).

At present, we replace the compression DCT method by the wavelets by computing the coefficients (L, H) which will be a soft threshold rather than quantified and coded (Maniccam 2001). The results are better than those obtained with the DCT. The results are illustrated in Figures 2 and 3 with $L = 8 \times 8$ and $N = 64$.

3.1 *Effect of the frame length*

Furthermore, we studied the effect of the size of the block L and the key length on the quality of the coded and deciphered image. In the third stage, simulations are made to test the robustness of the algorithm used according to various types of attack in particular during the transmission (Abdelmoula 2003).

Figures 4, 5, 6 and 7 represent the computing times in function of the segmental length L of the image and the ciphering length key. We notice that the computing time of ciphering/deciphering increases considerably with the two factors (block size and key length). A compromise is necessary to optimize the parameters and the ciphering quality. The simulation results give an optimum for L = 8 to 12 pixels as length of blocks and N = 1024 bits for the key.

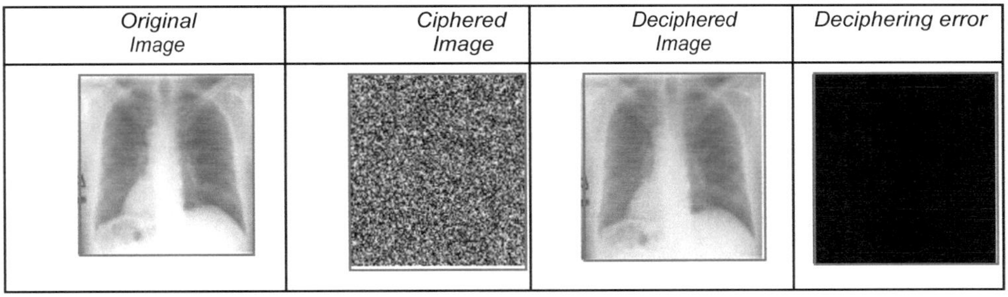

Original Image	Ciphered Image	Deciphered Image	Deciphering error

Figure 2.   Original and ciphered images.

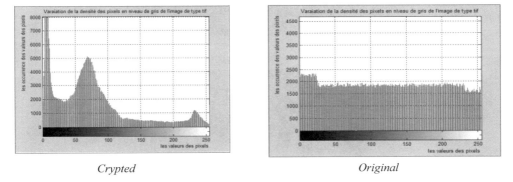

Crypted   Original

Figure 3.   Histograms of the original and ciphered images.

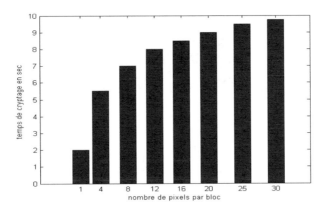

Figure 4.   Ciphering time vs frame length.

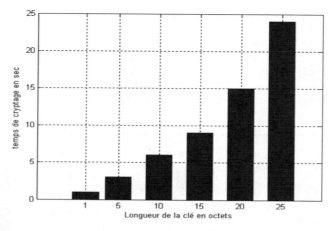

Figure 5.   Ciphering time vs key length.

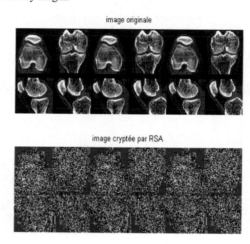

Figure 6.   Original and ciphered images for a bloc ($L = 6$ pixels).

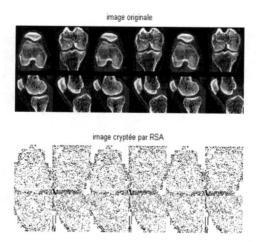

Figure 7.   Original and ciphered images for a bloc ($L = 8$ pixels)

Ciphering key (n,e) : (n :1024 bit)

n =
4464520673169410726480239794956030965579352068021207076088209357425564025906531634355260562664093661119470409675249431667900463823296207253792609956028537514413558435306822695613609213298836509206980179623731697971378941054409584486766856195773661880742677579589620491056145467464963831846362807192417733353359

e = 28669950953321621413

the ciphered message is C : c= m^e mod n =

c =
3128881659008038198849808727193317141771496772398881908448074662939711762406828968700618734083891475678911717704445717527141740728323972009268929365521660854890170145765805006966468159712100389499923340929349127172666129074136126189683248966126083900016621906703087370149810554074161861228622169732732487198042087627785731821654680260459538648107338763466417556874615907579431391106858007688854846414082020836641659792198093923226783637900348725016722047161229236219182577105627519094106797764091757593294643041155875208731647390520261605661587949028920239190229723782299990972985362014847053610199817915771928152503

Deciphering key (n,d) :

d =
4026723641123965394847333542454464801599875023543587974859999262881792880746722861331107051348288149506638875127159739781559578870075192158907550901477060438654660170038153885897326371022324336137746218717060617899535174733401997957490807693523040747025096043829418174975320950260628819869772466973903139879

Deciphered message is : out= c^d mod n =

out =
!('))+*&" ---- #&(,14<:99862-0++146?JUKA>BC=8552.+-27@CB@=>DJKHC?=:7487643210,-/13455654554201343126:IHC?BFE@@=;>CFC@54334

out =double(out) = bloc décrypté

033040039041041043042038034032031028029030030031031030029030032035038040044049052060058057057056054050045048048043043049052054063074085075065062066067061056053053050046043045050055064067066064061062068074075072067063061058055052056055054052051050049048044045047049051052053053054053052053053052050048049051052051049050050054058073072067063066070069064064061059062067070067064053052051051052

bloc image original : I

033040039041041043042038034032031028029030030031031030029030032035038040044049052060058057057056054050045048048043043049052054063074085075065062066067061056053053050046043045050055064067066064061062068074075072067063061058055052056055054052051050049048044045047049051052053053054053052053053052050048049051052051049050050054058073072067063066070069064064061059062067070067064053052051051052

Figure 8.   Example of an implementation of the RSA algorithm.

3.2   *Effect of the bloc length*

In this section, we can observe in Figures 6 and 7 that a length bloc under 8 pixels is not reliable.

Figure 8 shows the results of an example of the numerical results of an implementation of the RSA algorithm.

These results are confirmed by other research studies in reference such as Dhande (2015), and Puech (2004).

4 CONCLUSION

In this paper, we implemented an algorithm of encoding and ciphering of a digital image based on the R.S.A algorithm and intended for medical applications. This interface was implanted using MATLAB. It is based on the pre-processing of the original image based on a JPEG compression followed by a selective encoding quantification.

The obtained results demonstrate the importance of an optimization of two parameters: the segmented block length and the key length. Thus, we studied the effects of variation of these parameters on the deciphered image quality as well as on the processing time and the flow transmission.

REFERENCES

Abdelmoula M, Elloumi M, Kamoun L. 2003. "Nouveau schéma de crypto-compression des images médicales" *Revue Informatique, science de l'information et bibliothéconomie, RIST*, Vol. 13 No 2.

J.C. Borie, W. Puech, M. Dumas, 2002. Encrypted Medical Images for Secure Tranfer". International Conference on Diagnostic *Imaging and Analysis ICDIA 2002*, Shanghai. Vol 1, pp. 250–255.

C.C. Chang, M.S. Hwang, et T-S Chen, 2001. "A new encryption algorithm for image cryptosystems". *The Journal of Systems and Software*, vol 58, pp. 83–91.

Dhande M, A. Mudaliyar, V. Pandey, 2015. Secure Image Transmission using Cipher Block Chaining Mode & Visual Steganography. *International Journal of Science and Research*, vol 6.

S.S. Maniccam, N.G.Bourbakis, 20011. "Lossless image compression and encryption using SCAN". *Pattern Recognition, vol 34*, pp. 1229–1245.

Obo, Li, Jason Knipe, Howaed Cheng, 1997. "Image compression and encryption using tree structures". *Pattern Recognition Letters*, pp. 1253–1259.

Puech W, Borie J.C Dumas M, 2004. "Crypto-sompression system for Secure Tranfer of Medical Images". MEDSIP'04: *2nd Medical Image and Signal Processing*.

R. L. Rivest, A. Shamir and L. Adleman, 1978. "A method for obtaining digital signatures and public-key cryptosystems", *Communications of the ACM*, Vol 21, issue 2, pp. 120–126.

Innovative and Intelligent Technology-Based Services for Smart Environments – Ben Slama et al (eds)
© 2021 Taylor & Francis Group, London, ISBN 978-1-032-02030-3

Related review of the smart metering in a smart home and smart grid applications

M. Jraidi & A. Cherif
ATSSEE Laboratory, Science Faculty of Tunis, Tunisia

ABSTRACT: Smart metering is one of the most important infrastructures of the Smart Grid. It is based on intelligent measurement using smart meters, wireless sensor networks, remote networks for the bidirectional exchange of measurement data and the supervision center which allows the control, the treatment and the intelligent management of measurement data.

Smart metering is designed to allow the consumer to pilot his level of consumption in order to reduce it by the intelligent control of electronic devices of his home automation whose objective is to economize the use of electricity within a smart home. On the other hand, it is deployed by power service utilities through the advanced measurement infrastructure provided by smart readers to provide remote monitoring of large-scale consumers in a reliable and efficient manner. It works without reading errors due to the displacement of staff to view consumption levels from traditional meters and without using an estimation of consumption level from the history. It allows service and energy operators to manage the supply in a more reliable and intelligent way according to the balance between autonomous production and consumption and according to the geographical area which improves the services of the grid intelligent in the context of meeting the requirements of users and ensuring the safety of the entire electrical system.

Keywords: Smart metering, smart meters, smart home, smart grids, devices smart control, energy smart management, sensor networks

1 INTRODUCTION

In view of the shortage of electricity generated by conventional energy resources and the increase in electricity needs, which continue to grow, the use of renewable energy resources has become a necessity to address these limitations and to manage the growth of individual and industrial consumption.

The deployment of the smart grid to integrate renewable energy sources in an automated and intelligent way have become the most recent topics thanks to their major contributions in meeting energy demands and also their crucial role to opposing the scarcity of conventional resources by insisting on autonomous electric generation and by the minimizing of the use of traditional energies by resorting to the management of the conventional supply only for passive consumers and according to their needs and the limitations of their self-generation without causing network stability problems and without performing harmonics and incurring online electricity losses.

We can mention among the research studies that they are interested in the improvement of the smart grid, the approach of the cooperative communication that is based on the priority for the bidirectional consumption data aggregation via wireless sensor networks (Matta 2014) that focuses on the data processing reduces the cost of transmission and improves the response time to critical situations whose objective is to improve the service of response to demand. Another approach is the geographic protocol "Greedy" with a columniation and optimization based recovery strategy (Rekik 2016) which is the geographic routing protocol using the distribution of data to ensure the delivery of packets and the transfer of data to all participants. We also found that Net Metering (Seme 2017), which is the conventional electricity measurement approach transmitted to the consumer and the self-generated one, also returns the consumption data and measures the on-line losses between the energy emitted by the conventional or renewable resources and energy received at the level of the consumer. However, it does not contribute to the improvement of autonomous production or minimize consumption.

Other approaches are focused on measuring efficiency of grid-connected photovoltaic systems such as the measurement of the production performance under several climate factors and the measurement of the level of online losses at the production process level. Photovoltaic energy does not contribute to the improvement of the rate of energy productivity (Dabou 2016).

There is also the approach of the dynamic pricing strategies which are methods of voluntary control exercised by the consumer himself to reduce his consumption during peak periods in order to reduce the operational constraints of the system resulting in these peak periods of consumption.

We will treat in our approach the topic of smart measurement in the context of the smart home for the control of home automation devices to ensure the saving of energy consumed on a large scale in an automated way.

2 METHODS AND MATERIALS

Our remote energy measurement approach uses the advanced measurement infrastructure that monitors the data of the various meters integrated inside of the detailed consumer's smart homes, the advanced measurement infrastructure uses also the wireless communication capacities to ensure the two-way exchange of data between consumers and energy utilities providers. These various communicating tools will be integrated into a prototype of smart measurement for a smart home then for a residential city to take the necessary consumption measurements of different consumers that belong to the smart grid so that it will be simulated by simulators and supervised by our smart metering application that returns the consumption measurement data in order to allow suppliers and insurance service managers to ensure the balance between supply and demand on the one hand and to allow consumers to minimize their consumption by remote control of their home automation devices on the other hand.

2.1 *The smart metering process*

The operating basis of the energy smart metering process for the remote control of electronic devices allows their remote activation or deactivation with the framework of the reduction of consumption levels is as follows:

The current from the Alternating Current (AC) line is drawn for the circulation through a fuse to prevent damage of the circuit and the Arduino card during an accidental short circuit.

The Alternating Current (AC) power line is divided into two parts:

- A part for the charge through the Alternating Current Sensor (ACS712).
- Another part by the power supply module of 230V AC/5V DC.

The 5V power supply module gives power to the Arduino microcontroller, the esp8266 wifi card, the alternating current sensor (ACS712), the secure digital (SD) card module and the liquid card display (LCD). The alternating current passing through the load is detected by the current sensor module (ACS712) and transmitted to the analog pin (A0) of the Arduino card. Once the analog input is given to Arduino, the power/energy measurement is done by Arduino sketch.

The power and energy calculated by the Arduino board are displayed on an LCD display and will be sent to the server of the blynk application by the esp8266 wifi card to create a remote database that allows us to track our consumption in real time and on a large scale as well as it will also be recorded on an SD card to create a local database.

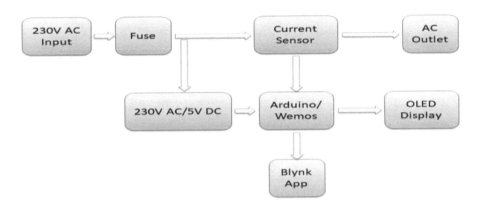

Figure 1.   Designing a diagram of a smart meter.

2.2   *The demonstration of our application of intelligent measurement and smart home automation management for the energy saving*

The smart measurement application is designed to measure the consumption of detailed users in a smart home. That is why we will design a prototype of a "smart home" that aims to provide simple and intuitive solutions to the problems of saving energy, safety and user comfort by centralizing the control of different electronic equipment. Our smart home prototype is to automate or to control remotely anything that can belong to the electronic components to allow the autonomous control of person's environment with the objective of minimizing its electricity consumption

Figure 2.   Smart home prototype with smart meter.

The figure 3 shows our smart home prototype that it integrates the smart meter to measure the electricity consumed during the activation of electronic devices within the smart home. It also demonstrates our smart home prototype that is focused on remote control of home components,

Figure 3. Demonstration of our remote door opener and remote light activation application.

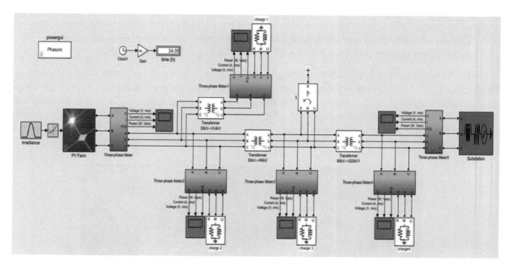

Figure 4. Model of a decentralized photovoltaic station using MATLAB/Simulink.

especially the automatic door opening and immediate activation of light bulbs within the smart home.

Similarly, these components will be deactivated in an automated manner upon the exit of a user by exercising the necessary commands for the closing of the door and the remote deactivation of the luminous bulbs.

2.3 Simulation of the smart grid architecture

Our intelligent measurement application will also be integrated into the smart grid to monitor the consumption data from conventional or renewable resources and to do this integration,;e will design the smart grid. The decentralized photovoltaic stations are equipped with the photovoltaic generator and the booster chopper to increase the voltage produced by Photovoltaic. The DC/DC converter will be equipped with MPPT technology (Maximum Power Point Tracking) to search

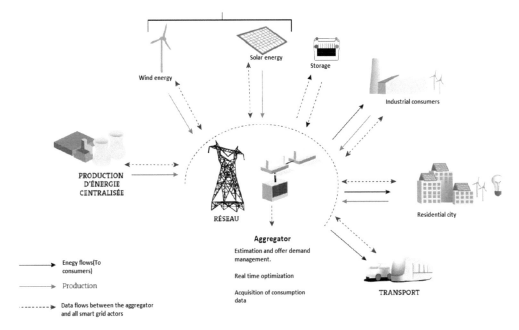

Figure 5. Visualization of smart grid features.

for the maximum power produced by an inverter to convert the DC voltage into a three-phase AC voltage to integrate it into the power grid.

The infrastructure needed to implement our smart metering application in the smart grids is composed of the following elements:

- Smart meters that are enabled by remote access networks and communicating digital sensors.
- Supervision and control centers: Data centers that allows monitoring, processing and decision making in place of old utilities.
- Automated actuators acting remotely.
- High performance sensors associated with existing devices.
- Bidirectional protection devices.
- Energy storage devices.

Our application is designed to measure the consumption that a consumer uses in detail. It can be applied to all renewable energies using smart meters, wireless communication networks, sensor networks that are activated by wireless communication networks and control and supervision centers to manage consumption and production as shown in the Figure 5.

3 MODELLING AND TOOLS

3.1 *Detailed study of our approach*

Our approach of the "Smart measurement and management system" is inspired by Net Metering (Seme 2017) which is a reliable method to measure the energy transmitted and the energy consumed to qualify online losses and the "Advanced Metering Management" (Colak 2016) which is the dynamic control approach based on automatic load data control and communication interfaces. Our approach to the smart energy management and metering system is based on the combination of Net Metering and Advanced Metering Management in a single Protocol called the "Smart

Measurements and Management System SMMS." This protocol is designed to ensure the functionality of the two approaches combined together by taking their advantages and minimizing their limitations while improving their results and offering other intelligent services and applications to enable consumers to consult their consumption on a large scale and in real time by an energy mix or by electronical devices with their comparisons with their neighbors in the same neighborhood network or by their comparison of their levels of consumption by geographical area while sending to them all the potential alerts and predictive notifications of power cuts, voltage drops, power consumption limits, extreme peak periods and potential failures that can occur and influence the user's electronic components. These alerts must be sent to make consumers aware of taking precautions by intentionally disabling nearby or remotely their electronic devices to improve life duration in case of sudden power failure and deploy other forms of self-generated energy stored in their storage media to catalyze the electricity lack and to meet the needs. Our approach is also interested in the monitoring of user data by services and energy utilities that they control, supervise, monitor and manage data on a large scale and in the real time by geographical area or by the level of energy autonomy while providing the conventional load to users whose energy needs exceed their self-generation levels.

The traditional system of energy consumption must be emerged and developed to become an intelligent system that uses information and communication technologies involving an embedded system and remote communication technologies to develop a new method of adjustable energy consumption which is controlled and organized. To do this, we dedicate ourselves to the implementation of a website to pilot the advanced measurement infrastructure by familiarizing it with the environment of smart grids to visualize the user interface of each authorized client being registered in the measurement services.

The various components of the advanced measurement system contain intelligent measuring equipment that is smart meters located at the level of each consumer and that allow two-way communication with the service provider in terms of transforming the electrical network into a complex information system called "Smart Grids." The transmission of data to the advanced measurement infrastructure is done by the energy gateway that it is an interface between the energy consumers and the system controllers and that allows the user to control the activation, deactivation and the operation of home automation network applications to collect real-time energy consumption data from smart meters and to provide a link with service providers enabling them to control the system on a large scale.

The energy gateway transmits the data to the advanced measurement infrastructure that ensures the control and two-way communication of this consumption data with the service and energy provider through the measurement and advanced remote management center. Meter Data Management (MDM) validates this data and estimates future consumption while asking consumers to reduce their consumption through the use of their smart devices to control their consumption on a large scale and in real time.

4 RESULTS AND DISCUSSION

4.1 *Results of user's consumption by different criteria*

After all the measurements are made for the creation of our consumption database, we will process the collected information to have detailed information about the grid from which users belonging to the grid can view and control their consumptions. Figure 6 shows the actual monthly consumption of the various consumers belonging to the smart grid, having smart homes endowed with intelligent management applications and belonging to different geographical areas, this consumption database is provided by the company of electricity and gas "STEG" to help us speed up our measurement process.

The Figure 7 shows the average daily consumption by home activity sectors via measuring the consumption in function of each hourly interval throughout a day and by visualizing the level of consumption marked by the various activities carried out by the consumer and demanding energy.

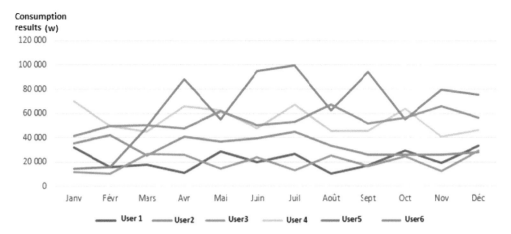

Figure 6.   Power consumption interface of the different users belonging to smart grids.

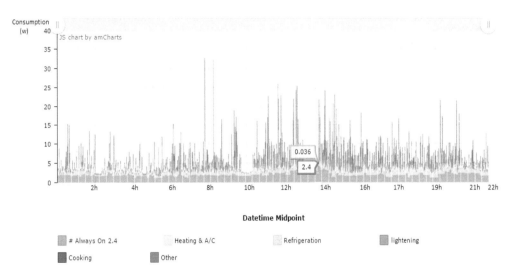

Figure 7.   Power consumption interface by activity sector.

The Figure 8 represents the power consumption of the conventional energy supplied from the electricity company and the gas instead of the fossil sources in the production of electricity and the consumption of self-generated energy coming from renewable sources.

4.2   *Discussion on the contribution of smart metering and home automation control in smart grids improvement*

Our approach supports several services and applications that are improving the performance of the smart grid, such as the support of the business process and allowing consumers to use their smart devices to manage autonomously their smart home automation business in the context of reducing their consumption level and while opposing the anomalies that they can appear in the behavior of the network. These applications help to improve several features of the smart grid that we distinguish.

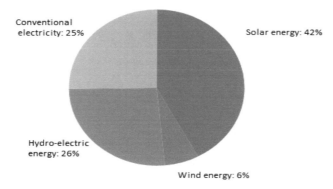

Figure 8. Power consumption interface by conventional and renewable energy source.

4.3 *Applications for advanced measurement infrastructure*

They are supported by the advanced measurement infrastructure that integrates the smart meters they are associated with with two-way communication technologies, these meters ensure the detection of the power of electronic devices, and support via our smart metering the measurement of electrical flows, the control and management of bidirectional flows of data and electricity also allow consumers to consult and track their consumption in real time and on a large scale to control their consumption during peak periods. These advanced measurement applications ensure the reliability, the speed and the accuracy of the consumption data to generate billing statements based on actual consumption based on undeniable measurement data and not estimated on historical and technical bases, these services offered allow consumers to follow their consumption profiles in real time and not only on periodic periods. They also allow them to consult their history, they allow them to reduce their consumption during peak periods by controlling and using remote driving of their homes appliances to reduce their consumptions and their billings.

4.4 *Distributed power grid management applications*

These applications are dedicated to operators who use smart metering for distributed automation-based distribution network management that provides real-time data collection and control of remote grid components, command support and decision support automated for the detection of anomalies and ensure the control and monitoring of the entire grid.

4.5 *Large scale network management applications*

They are dedicated to real-time monitoring of network behaviors and large-scale monitoring of the production and consumption data of each entity belonging to the network. These applications use smart metering to collect data from different entities in order to control the network and prevent it from any potential anomalies that they can prevent themselves as well as reduce the risk of disturbances. The technologies implemented for the control of the network that support the functionalities of our information system and our supervisor are the system of control, control and extended protection, these technologies are dedicated to the control and the piloting in order to protect the network.

5 CONCLUSION

This paper focuses on the application of smart metering as it provides large-scale and real-time data collection, which is designed for the measurement and the remote management of detail consumption and is done by the execution of various instructions such as the current and the voltage which are supervised by the control center using sensors and remote management systems that they ensure the control of all the instructions and parameters involved in the smart home as well as in the smart grid. These parameters and control instructions provide power, data flow and power control in a coordinated way.

REFERENCES

Bose K Artificial Intelligence Techniques in Smart Grid and Renewable Energy Systems — Some Example Applications, Proceedings of the IEEE | *Vol. 105, No. 11, November 2017.*

Adnane Cherif Intgration of renewable energy ressources for the production of decentralized energy, Proceedings of the JCGE'19 workshop Tunisia, 2019.

Ilhami Colak, A survey on the critical issues in smart grid technologies, Renewable and Sustainable Energy Reviews 54, 2016.

Rachid Dabou. Monitoring and performance analysis of grid connected photovoltaic under different climatic conditions in south Algeria, energy management and conversion, PHD 2016.

Nathalie Matta Decentralized management data of sensor networks in the contexte of the smart grids, PHD Thesis report, 2014.

Mouna Rekik Geographic routing protocol for the deployement of virtual power plant within the smart grids, Sustainable cities and society, elseiver, 2016.

Mouna Rekik Geographic greedy routing with ACO based void handling techniques, international journal of sensor networks, Inderscience, 2017.

Sebastian Seme Smart grids and net metering for photovoltaic systems, Institute of Electrical and Electronics Engineers, 2017.

Section IV: Digital marketing, smart tourism

Innovative and Intelligent Technology-Based Services for Smart Environments – Ben Slama et al (eds)
© 2021 Taylor & Francis Group, London, ISBN 978-1-032-02030-3

Facebook's role in promoting Tunisian tourism in Algeria

I. Boudraa, B.I. Nor El Houda & F.Z. Dehbi
University of Oran 2, Algeria

ABSTRACT: Nowadays, Social networking has become the new social media, this media is witnessing a wide movement of development, it is transforming from just a simple mean of communication into an integrated media tool that affects users in terms of decisions and preferences, as these sites have contributed into activating the sharing of interests and activities and also have certain influence and advocating Interact with new ideas and trends. E-tourism marketing in general and marketing via Facebook in particular is one of the modern and effective promotional methods that contribute onto influencing the conduct of tourists and their direction, especially with the huge rise that Facebook has achieved globally during the last decade. Based on the foregoing, this research paper addresses the impact of Facebook on promoting the tourist destination in Tunisia inside Algeria by relying on a questionnaire prepared and distributed to more than 50 Algerian tourism agencies active in the field of collective or individual organized trips to Tunisia, and by following the posts of its pages through Facebook and see the posts And likes and exchange pictures and videos about the tourism ingredients in Tunisia, and what these pages provide individually to introduce the Tunisian destination and its great tourism potential. For that, this research paper was divided alongside my theory deals with Facebook and its importance in marketing tourism and an applied aspect of his role in promoting Facebook for Tunisian tourism from Algerian tourism agencies.

*Innovative and Intelligent Technology-Based Services for Smart
Environments – Ben Slama et al (eds)*
© *2021 Taylor & Francis Group, London, ISBN 978-1-032-02030-3*

The role of big data analysis in improving network-based intellectual capital: The perspective of Social Network Analysis (SNA)

Fatima Zohra Dehbi
Oran2 University, Oran, Algeria

ABSTRACT: Intellectual Capital has attracted increasing attention and i widely considered a vital part of knowledge management. However, despite highlighting the importance of interactions between the network community and the knowledge society, because of the role of ICT infrastructure. The systematic description of network based-intellectual capital configurations has been scarce in the social network analysis perspective. This research seeks to contribute to Intellectual Capital (IC) research by incorporating big data analysis as an innovative and promising research tool providing a way to discover and capture knowledge from database systems, by data scientists and IT managers. Through the methodological process model for managing and analyzing big data within social network analysis (SNA), the findings demonstrated that the network-based Intellectual capital components must be interacted within the organizational network to facilitate the flow and exchange of information and extract new knowledge.

Keywords: Network-Based Intellectual Capital, Big Data Analysis, Social Network Analysis (SNA), Data Scientist, Discover New Knowledge.

1 INTRODUCTION

The development of ICT or NIT facilitates the emergence of the knowledge transfer and acquisition process. Furthermore, the widespread use of ICT has dramatically changed the diversity and structures of people's and organizations' relationships and resulting in the rise of a "network society, which generates network organizations kisielnicki and Sobolewska (2019). Organizational network analysis has recently attracted the researcher's attention Cross et al. (2010). With advances in the knowledge society and network community, researchers have acknowledged that the development of knowledge network communities is due to the organization's intranets and external knowledge sources Sarka et al. (2019). Consequently, the increased use of Network IT within the social networks makes information and knowledge easily flow. Thus, the network needs technology infrastructure, absorption platforms (Sheng 2019), and large databases for storing, analyzing, and finally extracting new knowledge Ekambaram et al. (2018). Maravilhas and Martins (2019, 355) maintains that in many organizations, a Chief Knowledge Officer (CKO) exists, and mediates all the programs between employees and the board of the organization. Generally, these processes could be realized through the contribution of knowledge workers especially KM analysts and data scientists. Knowledge networks are social systems. Thus, the creation of such networks is important for capturing knowledge and information, with redistributing them after analysis, consistent with the idea of Barão et al. (2017), that describes an organization as a social network comprising many networks and organizational memories which usually distribute the information throughout the network.

Following in this vein, the importance of this research lies in clarifying the role information communication technology plays as a tool for sharing and creating knowledge in organizations that use their technological means strongly.

DOI 10.1201/9781003181545-36

Despite the clear connection between ICT and Intellectual Capital, the comprehension of its impact on achieving accumulated organization data in database systems and its link with intellectual capital have been overlooked by pioneering scholars of the knowledge management area. Accordingly, there is a need to comprehend the literature's state-of-the-art regarding the potential of big data analysis in the linking of data and providing opportunities to extract new knowledge through the strategic role of social network analysis. In addition, research focuses on examining the role of data scientists and prove that structural capital is the most important component of intellectual capital in network organizations.

This research aims to advance the author's personal viewpoint and understanding of the analysis of big data from social network analysis, to discover and redistribute information and new knowledge to improve the stock of knowledge components of Network Based-Intellectual capital. In this respect, the author argues that big data analysis supports the emerging role of Network-Based Intellectual Capital as a critical resource for network organizations. Therefore, based on the above research gaps, this research answers the following question: "to what extent does the analysis of big data contribute to improving intellectual capital based on the network, from the perspective of social network analysis?"

2 INTELLECTUAL CAPITAL DIMENSIONS AS ANTECEDENTS FOR NETWORK ORGANIZATIONS

In recent decades, management literature has used the concept of Intellectual Capital (IC) to understand how knowledge stock of Intellectual capital functions as a key value-creating asset for organizations. In essence, Intellectual Capital represents the sum of all knowledge an organization can control in the process of directing business Youndt et al. (2004). Although various definitions of IC were presented by several researchers in the field, while the three-dimensional Intellectual Capital, is widely accepted and shared by most pioneering scholars, which are human capital (HC), relational capital (RC) and structural capital (SC) (Bontis 1998; Edvinsson and Sullivan 1996; Subramaniam & Youdnt 2005). On the one hand, Edvinsson and Sullivan (1996) contended that SC provides an environment that encourages human capital to create and use its knowledge. On the other hand, Roos & Roos (1997) found that human capital refers to an organization's employees and their attributes, such as knowledge and experience. However, researchers in the field believe that human capital is the most considerable component of intellectual capital because an organization can't accomplish innovation without it. Relational capital (RC) refers to interactions among an organization and its stakeholders, including the resulting networks of relationships the company maintains with external partners (Bontis 1998). Furthermore, with the widespread use of technological applications and the advent of the digital era, Secundo et al. (2020) emphasized that these potentialities increasingly enhance knowledge sharing and makes IC affect the innovation in learning. Network organizations face environmental challenges including rapid change in new technologies, because of this development.

Increasingly available ICT (as a component of structural capital) may offer opportunities to IC. According to, Stephens et al. (2016), analyzing structural processes that form the employees' network structure allows to the platform captures real-time digital trace data from all employee participation within a global IT organization. A recent study by Buenechea-Elberdin et al. (2018) has found that Structural Capital (SC) supports the accumulation of knowledge resources in companies' structures, databases, and information systems, and the knowledge embedded in internal and external networks of relationships because knowledge derives from data and information. Structural Capital can function as the collective knowledge stock and infrastructure that facilitates collaboration between employees and customers Hussinki et al. (2019). Thus, structural capital contributes to shaping the network-based intellectual capital.

3 BIG DATA ANALYSIS

Big data is about a large amount of data that requires a new way to process it. Traditional databases cannot adequately address issues related to collecting, storing, processing, or analyzing huge data sets Ekambaram et al. (2018). Lately, data is becoming a new form of an innovative source of value, and it's considered a powerful strategic process to channel manage and grow the business. Accordingly, before investing heavily in technology and infrastructures for big data, managers must start with checking to validate whether big data may provide value for an organization. (Corea 2019). The term 'Big Data' was originally used to describe large collections of data. Big data is at the core of data-driven research where research goals are driven by the preliminary analysis of patterns existent of data. Such data can be found in most information systems platforms Asamoah and Sharda (2019). From the perspective of the process, big data refers to the infrastructure and technologies that the organizations use to collect, store, and analyses various types of data Heang and Khan (2015).

King (2017) said "Big Data is not about the data," the incredible advances in statistical methods, new computer science and original theories in the field of application enable the extraction and discovery of knowledge from the data and this is the real value of big data. We admit the challenge is how to extract knowledge from the vast volume of data collected, especially from the dataset of network organization. Clearly, this large-volume data lays some challenges for the data analysis across, and to achieve new knowledge. Safhi et al. (2019) have claimed that Big Data knowledge extraction is the process of turning that data into usable information. The exponential growth of data has initiated a myriad of new opportunities and made data become the most valuable raw material of production for many organizations. Data mining represents the core of the knowledge discovery as shown in figure 1.

In this framework, it is the turn of data scientists and analysts to describe the quality of data that can be used. Therefore, it has been suggested that they identify principle classes to describe the quality of a particularly linked dataset data publishers can use which to refine and improve their datasets Safhi et al. (2019).

3.1 Social Network Analysis

Social network research originated from social science. The concept is mainly discussed concerning the concept of social capital, which is becoming a hot topic in intellectual capital studies after Naphiat & Ghoshal (1998, 243) defined it as "the sum of the actual and potential resources embedded within, available through and derived from the network of relationships possessed by an individual or social unit." Social networks emphasize the network relationships among individuals, teams, organizations, and communities that possess social resources (Inkpen & Tsang 2005).

An organization is a social network that comprises many smaller networks and typically the knowledge is dispelled throughout the network. As a result, the of ICT and knowledge broker/online intermediaries in these spaces has increased. Networks are useful for sharing knowledge over the boundaries of an organization. An organizational unit's network position and absorptive capacity represents its ability to leverage useful knowledge residing in other parts of its organization

Figure 1. **Source:** Safhi et al. (2019).

(Tsai 2001). Similarly, direct contact among employees from different organizations leads to more efficient knowledge sharing and therefore a higher absorptive capacity (Schmidt 2010). The knowledge broker the person tasked with the coordination of the society has the greatest impact on the level of collaboration among members Cross et al. (2010). The best knowledge brokers increase membership and encourage frequent participation. Indeed, in knowledge networks, correlations are made through the intentional effort by organizations and/or individuals to share and get knowledge resources to the benefit of network participants Vicente et al. (2011). Organizations should be aware that these networks can be an important way to access and share knowledge Janowicz-panjaitan & Noorderhaven (2008).

4 BIG DATA ANALYSIS AS ANTECEDENTS FOR FOSTERING NETWORK-BASED INTELLECTUAL CAPITAL

The ability to collaborate effectively is an important competence for network organizations. The networks called researchers' attentions to studying markets as systems of networks and acknowledges that inter-organizational activities (Marchiori & Franco 2019; Nawinna & Venable 2019; Sheng 2019) take place within the context of network relationships.

The development of the digital revolution accelerates the knowledge flow within the network community. The utilization of social network analysis in network-based Intellectual capital research is far from exhausted. For this reason, the author's personal viewpoint is based on a framework that defines enhancing the value of intellectual capital components, by extracting new knowledge and sharing it from big data analysis depending on social network analysis.

The structural model of modern companies that embraces networks, platforms, and crowdsourcing spur individuals' wisdom and abilities (Studies 2019). Herein, regulatory networks take different forms, which may be horizontal or vertical, formed between competitors, customers, suppliers, local or international partners Álvarez et al. (2009). According to Studies 2019, 66, "the database is the core software of data organization, storage, and management of the computer system. Its development includes, network and hierarchical databases, relational databases." Knowledge discovery capacity in network organization refers to transforming data into information and information into knowledge. Through contextualization, categorization, measurement, correction and condensation, structured and contextualized data becomes information Davenport & Prusak (1998). Öner & Yüregir (2019, 61) defined knowledge discovery as "a significant extraction of previously unknown and useful information from data. Data DM is considered as one part of the processes of knowledge discovering." Therefore, to gain a variety of knowledge, organizations need to engage the big data analysis to derive knowledge from data and information. Overall, we understand that the discovery of knowledge is an essential process fostering network-based intellectual capital, simultaneously require supporting organizational network analysis to discover and transfer knowledge quickly, and efficiently from big data.

As it is shown in figure 2, database technology roles are integrated with the knowledge network. Barão et al. (2017) clarified that the KM analysts and the IT Manager can extract knowledge from Big Data. The process impacts are illustrated on intellectual capital by enabling capturing organizational knowledge.

For more understanding of what the impacts are illustrated Big Data on intellectual capital components by network based-structural capital. The author argued that in parallel with knowledge management technology processes and brokers (e.g., data scientists, data analyst), it may be possible to discover new knowledge after analyzing the Big Data and storing the knowledge gathered from organizational networks that contributes to the enhancing of various components of intellectual capital and thus its value increases.

From the perspective of network-based human capital, the organization should adapt to the trend. (Widyaningrum 2016) found that learning processes lead knowledge workers to share their knowledge constantly from both internal and external sources. Furthermore, human resources would put more attention to the need for new experts such as data scientists and IT managers.

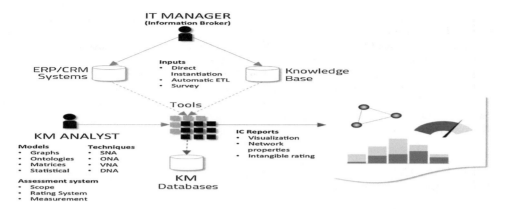

Figure 2. **Source**: Barão et al. (2017).

Furthermore, Madhavaram and Hunt's (2017) study showed that the structural dimension can facilitate information exchange with clients and cooperative behaviors. Admittedly, it is essential for intimate knowledge of customer requirements, wants and preferences. Galbraith 2014 also came to a similar conclusion, which emphasized that the network organizations must embed IT managers and data to execute this into decision processes that generate new understandings from big data, for setting customer priorities, making decisions on new product features that make the network organizations more successful.

Moreover, the rapid development of ICT has an important impact on the spread of knowledge. In this case, Shan et al. (2019) have measured the major influencing factors in the development of IT technology capabilities, to adapting to the changes in the market environment and customer needs. The results proved the importance of IT relationship resources includes the relationship between the company and the customer as well as the critical relationship between business and IT.

Against this backdrop and based on the study (Nahapiet & Goshal 1998; Inkpen & Tsang 2005), the configuration of a network structure determines the pattern of linkages among network members and knowledge transfer facilitated through network structure. Hence, the author argues network-based intellectual capital is generated through the exchange of knowledge, which links network members in the structural dimension of social networks. Incorporating intellectual capital in organization networks in research thus has the potential to produce a more comprehensive understanding of the role of social network analysis.

In line with this overview, the author defines network-based intellectual capital as: "construct network-based intellectual capital is the process of value creation and acquisition of knowledge for the benefit of network participants within social networks." The configuration of network-based intellectual capital implies two steps:

1) Increasing network interconnections. The interaction between individuals and organizations, as channels, allows the flow of data and information which can be easily stored in databases.
2) Employing a data set as a source of big data, and processing it through data capturing, monitoring and interpretation to discover new knowledge.

5 CONCLUSION

There is growing interest in understanding the big data phenomenon and its influence on the different environments of the organization. Nevertheless, several researchers of the intellectual capital area neglected to analyze the big data which represents the new knowledge stock of intellectual capital components within network society. This research compared the research themes proposed by different literature reviews, addressed the growing needs for knowledge extracted from big data and emphasized the importance of the social construction of data science and represented an attempt to make sense of understanding the processes of knowledge exchanges between the components of in-intellectual capital in network organization through the skills of data scientists, analysts and IT managers. The interactions between multiple resources allowed gain from different units of an organizational network in accumulating data and information. Therefore, data scientists contribute to extracting value from big data, which generates new knowledge.

The emergence of the concept of network-based intellectual capital faces a hyper-competitive business environment, which is digital, virtual and networked.

The analysis of big data from social network analysis contributes to IT managers' abilities to discover and redistribute new knowledge to foster knowledge stock of network based-intellectual capital components.

The database and platforms contribute by providing a ground for structural capital, to sharing knowledge across networks and, in turn, improving the network-based intellectual capital components.

In light of the emergence of networks and their emphasis on investing in acquiring an organization's relationships, organizational network capacity considers achieving and exploiting the technology to create value and advantages by facilitating knowledge exchange. Network-based intellectual capital management is being considered a central resource.

To summarize, the impact big data analysis has on improving network-based intellectual capital and inspiring network organizations to achieve, improve and strengthen their network-based intellectual capital in the digital economy, must pay attention to their structural capital, because the database is the core software of data organization. Using database technology is integrated with network communication, the organization needs professionals who are skilled with digital tools to work in teams. Such systems allow the transfer of knowledge resources between team members, either directly through mediated interactions, or indirectly by searching knowledge repositories such as databases.

All interactions and relationships between employees, customer relation management and structural capital need to harness various databases, and other types of data and process them, to uncover customers' influences on networks and to connect with customers.

REFERENCES

Álvarez, I, R Marin, and A Fonfría. 2009. "The Role of Networking in the Competitiveness of Firms." *Technological Forecasting and Social Change* 76 (3): 410–21. https://doi.org/10.1016/j.techfore.2008. 10.002.

Asamoah, D A, and R Sharda. 2019. "CRISP-ESNeP: Towards a Data-Driven Knowledge Discovery Process for Electronic Social Networks." *Journal of Decision Systems* 28 (4): 286–308. https://doi.org/10.1080/ 12460125.2019.1696614.

Barão, Al, J B de Vasconcelos, Ál Rocha, and R Pereira. 2017. "A Knowledge Management Approach to Capture Organizational Learning Networks." *International Journal of Information Management* 37 (6): 735–40. https://doi.org/10.1016/j.ijinfomgt.2017.07.013.

Bontis, Nick. 1998. "Intellectual Capital: An Exploratory Study That Develops Measures and Models." *Management Decision* 36 (2): 63–76.

Buenechea-Elberdin, Marta, Josune Sáenz, and Aino Kianto. 2018. "Knowledge Management Strategies, Intellectual Capital, and Innovation Performance: A Comparison between High- and Low-Tech Firms." *Journal of Knowledge Management* 22 (8): 1757–81. https://doi.org/10.1108/JKM-04-2017-0150.

Corea, F. 2019. "Introduction to Data." In *An Introduction to Data*, 1–5. Springer.

Cross, R L, J Singer, S Colella, R J Thomas, and Y Silverstone. 2010. *The Organizational Network Fieldbook: Best Practices, Techniques and Exercises to Drive Organizational Innovation and Performance*. John Wiley & Sons.

Davenport, T H, and L Prusak. 1998. *Working Knowledge: How Organizations Manage What They Know*. Boston: Harvard Business School Press.

Edvinsson, L, and P Sullivan. 1996. "Developing a Model for Managing Intellectual Capital." *European Management Journal* 14 (4): 356–64. https://doi.org/10.1016/0263-2373(96)00022-9.

Ekambaram, A, A Ø Sørensen, H Bull-berg, and O E Olsson, N. 2018. "The Role of Big Data and Knowledge Management in Improving Projects and and Project-Based Organizations." In *Procedia Computer Science*, 138:851–58. Elsevier B.V. https://doi.org/10.1016/j.procs.2018.10.111.

Galbraith, J R. 2014. "Organization Design Challenges Resulting from Big Data" 3 (1): 2–13. https://doi.org/10.7146/jod.8856.

Heang, J F, and H U Khan. 2015. "The Role of Internet Marketing in the Development of Agricultural Industry: A Case Study of China." *Journal of Internet Commerce* 14 (1): 65–113. https://doi.org/10.1080/15332861.2015.1011569.

Hussinki, H, A Kianto, M Vanhala, and P Ritala. 2019. "Happy Employees Make Happy Customers: The Role of Intellectual Capital in Supporting Sustainable Value Creation in Organizations." In *Intellectual Capital Management as a Driver of Sustainability*, 101–17. Springer.

Inkpen, A C., and E W.K. Tsang. 2005. "Social Capital, Networks, and Knowledge Transfer." *Academy of Management Review* 30 (1): 146–65. https://doi.org/10.5465/AMR.2005.15281445.

Janowicz-panjaitan, Martyna, and Niels G Noorderhaven. 2008. "Formal and Informal Interorganizational Learning within Strategic Alliances." *Research Policy* 37: 1337–55. https://doi.org/10.1016/j.respol.2008.04.025.

King, G. 2017. "Preface: Big Data Is Not About The Data!" 2017. https://gking.harvard.edu/files/gking/files/prefaceorbigdataisnotaboutthedata_1.

Kisielnicki, Jerzy, and Olga Sobolewska. 2019. "Knowledge Management and Innovation in Network Organizations: Emerging Research and Opportunities: Emerging Research and Opportunities." In, Tavana, Ma. Advances in Business Information Systems and Analytics. IGI Global. https://doi.org/10.4018/978-1-5225-5930-6.

Madhavaram, S, and S D Hunt. 2017. "Customizing Business-to-Business (B2B) Professional Services: The Role of Intellectual Capital and Internal Social Capital ☆" *Journal of Business Research* 74: 38–46. https://doi.org/10.1016/j.jbusres.2017.01.007.

Maravilhas, Sérgio, and Joberto Martins. 2019. "Strategic Knowledge Management a Digital Environment: Tacit and Explicit Knowledge in Fab Labs." *Journal of Business Research* 94 (August 2017): 353–59. https://doi.org/10.1016/j.jbusres.2018.01.061.

Marchiori, D, and M Franco. 2019. "Knowledge Transfer in the Context of Inter-Organizational Networks: Foundations and Intellectual Structures." *Journal of Innovation & Knowledge* 5 (2): 130–39.

Nahapiet, J, and S Goshal. 1998. "Social Capital, Intellectual Capital, and The Organizational Capital Advantage." *Academy of Management Review* 23 (2): 242–66.

Nawinna, Dasuni, and John R. Venable. 2019. "Effects of ICT-Enabled Social Capital on Inter-Organizational Relationships and Performance: Empirical Evidence from an Emerging Economy." *Information Technology for Development* 25 (1): 49–68. https://doi.org/10.1080/02681102.2018.1451979.

Öner, S C, and O H. Yüregir. 2019. *Optimizing Big Data Management and Industrial Systems With Intelligent Techniques*. https://doi.org/10.4018/978-1-5225-5137-9.

Roos, Göran, and Johan Roos. 1997. "Measuring Your Company's Intellectual Performance." *Long Range Planing* 30 (3): 413–26.

Safhi, H M, B Frikh, and B Ouhbi. 2019. "Assessing Reliability of of Big Data Knowledge Discovery Process." *Procedia Computer Science*. Elsevier B.V. https://doi.org/10.1016/j.procs.2019.01.005.

Sarka, P, P Heisig, Ni H.M. Caldwell, A M. Maier, and C Ipsen. 2019. "Future Research on Information Technology in Knowledge Management." *Knowledge and Process Management* 26 (3): 277–96. https://doi.org/10.1002/kpm.1601.

Schmidt, T. 2010. "Absorptive Capacity-One Size Fits All? A Firm-Level Analysis of Absorptive Capacity for Different Kinds of Knowledge." *Managerial and Decision Economics* 31 (1): 1–18. https://doi.org/10.1002/mde.1423.

Secundo, Gi, V Ndou, P Del, and G D P Pascale. 2020. "Sustainable Development, Intellectual Capital and Technology Policies: A Structured Literature Review and Future Research Agenda." *Technological Forecasting & Social Change* 153: 119917. https://doi.org/10.1016/j.techfore.2020.119917.

Shan, S, Y Luo, Y Zhou, and Y Wei. 2019. "Big Data Analysis Adaptation and Enterprises' Competitive Advantages: The Perspective of Dynamic Capability and Resource-Based Theories." *Technology Analysis and Strategic Management* 31 (4): 406–20. https://doi.org/10.1080/09537325.2018.1516866.

Sheng, Margaret L. 2019. "Foreign Tacit Knowledge and a Capabilities Perspective on MNEs' Product Innovativeness: Examining Source-Recipient Knowledge Absorption Platforms." *International Journal of Information Management* 44: 154–63. https://doi.org/10.1016/j.ijinfomgt.2018.10.008.

Stephens, Bryan, Wenhong Chen, and John Sibley Butler. 2016. "Bubbling Up the Good Ideas: A Two-Mode Network Analysis of an Intra-Organizational Idea Challenge." *Journal of Computer-Mediated Communication* 21 (3): 210–29. https://doi.org/10.1111/jcc4.12158.

Studies, Chinese Academy of Cyberspace. 2019. "Development of the World's Network Information Technology." In *World Internet Development Report 2017*, 55–87. Heidelberg: Springer, Berlin, Heidelberg. https://doi.org/10.1007/978-3-662-57524-6.

Subramaniam, Mohan, and Mark A. Youdnt. 2005. "The Influence of Intellectual Capital on the Types of Innovative Capabilities." *Academy of Management Journal* 48 (3): 450–63.

Tsai, Wenpin. 2001. "Knowledge Transfer in Intraorganizational Networks: Effects of Network Position and Absorptive Capacity on Business Unit Innovation and Performance." *Academy of Management Journal* 44 (5): 996–1004. https://doi.org/10.2307/3069443.

Vicente, J, P A. Balland, and O Brossard. 2011. "Getting into Networks and Clusters: Evidence from the Midi-Pyrenean Global Navigation Satellite Systems (GNSS) Collaboration Network." *Regional Studies* 45 (8): 1059–78. https://doi.org/10.1080/00343401003713340.

Widyaningrum, D T. 2016. "Using Big Data in Learning Organizations." In *3rd International Seminar and Conference on Learning Organization (ISCLO 2015) Using*. Atlantis Press.

Youndt, M A., M Subramaniam, and S A. Snell. 2004. "Intellectual Capital Profiles: An Examination of Investments and Returns." *Journal of Management Studies* 41 (2): 335–61. https://doi.org/10.1111/j.1467-6486.2004.00435.x.

Innovative and Intelligent Technology-Based Services for Smart Environments – Ben Slama et al (eds)
© 2021 Taylor & Francis Group, London, ISBN 978-1-032-02030-3

The mediating role of operational costs on financial performance of Sharia Banks in Indonesia

Ana Kadarningsih, Irene Rini Demi Pangestuti, Sugeng Wahyudi & Vicky Oktavia
Diponegoro University, Semarang, Indonesia

ABSTRACT: The objectives of this study are to determine the effects of the Capital Adequacy Ratio (CAR), Financing to Deposit Ratio (FDR) and Non-Performing Financing (NPF) on the Return on Assets (ROA) mediated by operational costs to operational income. The samples of this study are 12 Sharia Banks in Indonesia that published their financial reports to the Financial Services Authority in 2013 until 2019. The analysis technique used is regression analysis and path analysis. The results of this study are that Capital Adequacy Ratio has a positive significant effect on financial performance in Sharia Banks. Secondary results are that the Financing to Deposit Ratio and Non-Performing Financing have an insignificant, negative impact on financial performance in Sharia Banks. Our third results are that operational costs have a negative, significant impact on financial performance in Sharia Banks. Operational costs cannot mediate the Capital Adequacy Ratio, Financing Deposit Ratio and Non-Performing Financing on financial performance in Sharia Banks. The implication of this research is managers of Sharia Banks in Indonesia need to prioritize policies related to the Capital Adequacy Ratio and Operational Costs for profitability's increasing.

Keywords: Operational Costs, Financial Performance, Sharia Bank

1 INTRODUCTION

Based on data sourced from Susenas 2019 recorded at the Central Statistics Agency (BPS), the total population of Indonesia has reached 267 million. 232 million people (87.17%) of the Indonesian population are Muslim. Even though the majority of Indonesian people are Muslim, there are many challenges that Islamic banking must face in developing its activities, one of which is the lowering of public awareness related to Islamic banking (Wilantini & Dewi 2019). In early 2015, the Sharia Banking Department within the OJK stated that after experiencing relatively high growth (1.25%) in previous years, in 2013 the Islamic banking experienced a growth slowdown in line with slowing economic growth, and experienced a drastic decline of 1.5% in 2014 (Financial Services Authority 2015).

Research from Trisningtyas (2013) states that FDR and CAR have no effect on bank financial performance (ROA) while BOPO and NPF have an effect on ROA in Islamic banks. Tristiningtyas et al. (2013) also said that it is necessary to reduce NPF and BOPO to increase profitability in Islamic banks. The results of this study are different from Yunita (2014) which shows that FDR and CAR can affect ROA in positively, but BOPO and NPF have a negative influence on ROA. Ubaidillah (2016) provides research results that are slightly different from previous studies, where the results of testing the hypothesis show that NPF, BOPO and CAR can affect ROA in the negative significantly, while FDR has a positive, significant impact on ROA

DOI 10.1201/9781003181545-37

2 LITERATURE REVIEW AND HYPOTHESES DEVELOPMENT

2.1 The effect of CAR, FDR, and NPF on BOPO

Capital Adequacy Ratio (CAR) is a ratio to measure the bank's capital adequacy (Nahrawi 2017). According to BI regulation Number 10/15/PBI/2008 article 2 paragraph 1, it is stated that the minimum capital of banks is 8% of risk weighted assets (RWA). The previous studies describe that CAR has a negative significantly effect on BOPO (Puspitasari 2009). Migustin (2017) in his research said that CAR does not affect BOPO. So hypothesis 1 (H1) is:

CAR has a negative significantly impact on BOPO in Sharia Banks.

Financing to Deposit Ratio (FDR) is a ratio to define the liquidity of a bank (Riyadi & Yulianto 2014). The higher the FDR ratio, the lower the BOPO ratio (Ailiyah 2019). According to BI Circular Letter No. 6/23/DPNP/2004, a good FDR value is 80%. Research from Ayuningrum and Widyarti (2011) concluded that FDR had a positive, significant effect on BOPO. Hypothesis 2 (H2) can be formulated as below:

FDR has a negative significantly effect on BOPO in Sharia Banks

Non-Performing Financing (NPF) is a ratio used to measure failure in returns from financing channeled by banks (Poernamawatie 2009). Bank Indonesia Circular Letter No. 17/19/DPUM/2015 set a minimum NPF of 5% for Islamic Commercial Banks. Poernamawatie (2009) found the NPF variable has a positive effect on BOPO. However, research from Puspitasari (2009) states that NPF has a negative effect on BOPO. As a result of previous research, hypothesis 3 (H3) is:

NPF has a positive and significant effect on BOPO in Sharia Banks

2.2 The effect of CAR, FDR, NPF and BOPO on ROA

Operating Costs to Operating Income (BOPO) is often called the efficiency ratio which is used to measure the ability of bank management to control operating costs against operating income (Trisningtyas 2013). Bank Indonesia sets the ideal BOPO Ratio limit to be below 90%, because then the bank can be categorized as efficient in carrying out its operational activities. When a bank is in good health, the bank's performance will be better and the ROA will be higher (Hermawan & Fitria 2019). Previous studies state that BOPO has a negative and significant effect on ROA in Islamic Commercial Banks (Teshome et al. 2018; Ubaidillah 2016; Hermawan & Fitria 2019). Otherwise, BOPO has a positive and significant effect on ROA in Islamic Commercial Banks (Rizkika et al. 2017; Nanda et al. 2019).

Return on Assets (ROA) is a profitability ratio for financial performance that shows the company's capacity to earn a profit with all company assets. ROA is a comparison of Earning Before Interest and Tax (EBIT) to total assets (Harmono 2017). Previous studies found that there is a positive impact of CAR on ROA in Islamic commercial banks (Tarawneh et al. 2017; El Maude et al. 2017; Teshome et al. 2018). Another research result shows that the CAR variable has no effect on ROA (Ubaidillah 2016; Rizkika et al. 2017; Hermawan & Fitria 2019). Hafiz et al. (2019) found that CAR has an insignificant and negative impact on ROA.

The results of previous research are that the FDR variable has a significant influence on ROA in Islamic Commercial Banks (Ubaidillah 2016; El Maude et al. 2017; Tarawneh et al. 2017; Hermawan & Fitria 2019). Meanwhile, another result found that FDR has insignificant effect on ROA in Islamic Commercial Banks (Trisningtyas 2013; Rizkika et al. 2017, Hafiz et al. 2019). NPF has a negative significantly impact on ROA in Sharia Banks (Nahrawi 2017; Ratnawaty 2018; Hermawan & Fitria 2019; Ozili 2017). Hafiz et al. (2019) analyzes that NPF has an insignificant and negative effect on ROA in Islamic Commercial Banks, while Rizkika et al. (2017) stated that NPF has a significant and positive influence on ROA in Islamic Banks.

Based on previous studies, so there are another four hypotheses as below:

H4. CAR has a positive significantly effect on ROA in Sharia Banks

Table 1. Determination coefficient test results of first regression.

		Model Summary[b]		
Model	R	R Square	Adjusted R Square	Std. Error of the Estimate
1	.808[a]	.653	.640	.153429

a. Predictors: (Constant), NPF(X3), CAR(X1), FDR(X2)
b. Dependent Variable: BOPO(Y1)

Table 2. Path analysis with BOPO as the dependent variable.

		Coefficients[a]				
		Unstandardized Coefficients		Standardized Coefficients		
Model		B	Std. Error	Beta	t	Sig
1	(Constant)	.862	.027		32.288	.000
	CAR(X1)	−.122	.064	−.147	−1.920	.058
	FDR(X2)	.000	.000	.572	7.464	.000
	NPF(X3)	2.628	.261	.666	10.071	.000

a. Dependent Variable: BOPO(Y1)

H5. FDR has a positive significantly effect on ROA in Sharia Banks
H6. NPF has a negative significantly effect on ROA in Sharia Banks
H7. BOPO has a negative significantly effect on ROA in Sharia Banks

3 RESEARCH METHODS

This research used a descriptive quantitative research method with SPSS tools version 23. The type of data is secondary data, namely annual financial reports published by Islamic Commercial Banks registered in the Financial Services Authority from 2013–2019. Sampling was carried out in this study using a purposive sampling technique. The samples of this research are 12 Sharia Banks in Indonesia. The technique used in this research is regression analysis and path analysis. This study examined the effect of the CAR, FDR, and NPF variables on ROA with BOPO as an intervening variable. This research used partial testing (T test) and the coefficient of determination (R2) to analyze hypotheses. Then, the intervening test was performed using path analysis.

4 RESULTS AND DISCUSSION

The first regression equation is BOPO as a dependent variable which can be seen in Tables 1 and 2.
    The value of R Square found in Table 1 is 0.653. This explains that the contribution of the influence of CAR, FDR and NPF on BOPO is 65.3%, while 34.7% is the contribution of other variables outside this study.
    Based on the regression coefficient output in Table 2, it can be seen that the significance value of CAR (X1) = 0.058 is greater than 0.05, these results indicate that CAR has a insignificant effect on BOPO. So, H1 is rejected. While the significance value of the variable FDR (X2) = 0.000 and the variable NPF (X3) = 0.000 is smaller than 0.05. These results conclude that the FDR has positive

Table 3. Determination coefficient test results of second regression.

		Model Summary[b]		
Model	R	R Square	Adjusted R Square	Std. Error of the Estimate
1	.853[a]	.727	.713	.023775

a. Predictors: (Constant), BOPO(Y1), CAR(X1), NPF(X3), FDR(X2)
b. Dependent Variable: ROA(Y2)

Table 4. Path analysis with ROA as the dependent variable.

		Coefficients[a]					
		Unstandardized Coefficients			Standardized Coefficients		
Model		B	Std. Error		Beta	t	Sig
1	(Constant)	.119	.015			7.706	.000
	CAR(X1)	.047	.010		.327	4.674	.000
	FDR(X2)	−5.962E-6	.000		−.062	−.699	.487
	NPF(X3)	−.110	.061		−.161	−1.812	.074
	BOPO(Y1)	−.120	.017		−.690	−5.906	.000

a. Dependent Variable: ROA(Y2)

and significant effect on BOPO. So, H2 is rejected. NPF variables have positive and significant effects on BOPO. So, H3 is accepted.

The second regression equation is ROA as a dependent variable which can be observed in the Tables 3 and 4 below:

The value of R Square found in Table 3 is 0.727. This describes that the contribution of the influence of CAR, FDR, NPF and BOPO to ROA is 72.7%, while 27.3% is the contribution of other variables outside this study.

Based on the second regression on Table 4, it can be concluded that the significance value of CAR (X1) = 0.000, the variable FDR (X2) = 0.487, the variable NPF (X3) = 0.074 and the variable BOPO (Y1) = 0.000. These results conclude that CAR has a significant and positive impact on ROA, while NPF and FDR have an insignificant effect on ROA. So, H4 is accepted but H5 and H6 are rejected. BOPO has a negative and significant impact on ROA, so H7 is accepted. The results of two path analysis tests show that BOPO cannot mediate CAR, FDR and NPF to ROA.

5 CONCLUSION AND RECOMMENDATION

The partial test found that CAR has a significant and positive impact on ROA, which means that if CAR increases, the profitability of Sharia banks will increase too. FDR and NPF have negative and insignificant impacts on ROA. Therefore, FDR as the liquidity ratio and NPF as the level of bad credit shows that the level of bad credit does not have a major influence in measuring the profitability of Sharia banks. BOPO has a negative, significant impact on ROA, which means that the higher the BOPO will reduce the ROA generated by Sharia Banks. CAR and FDR do not impact BOPO, only NPF has positive and significant effect on BOPO. But BOPO cannot mediate the relationship between CAR, FDR and NPF on ROA.

It is recommended that Sharia banks increase penetration in the Islamic banking sector and increase public awareness regarding Sharia banking which is still low, for example by increasing cooperation in the MSME development sector which has a major contribution to the Indonesian economy through the provision of business credit.

Future researchers are expected to be able to find a different and wider population scope to provide a more specific picture of the effects of CAR, FDR, NPF and BOPO on ROA in Sharia Banks. Other variables are size, leverage and inflation with other research objects such as commercial banks, state-own banks or leasing companies.

REFERENCES

[1] Ayuningrum, Anggrainy Putri. Widyarti, Endang Tri (2011) The Effect of CAR, NPL, BOPO, NIM and LDR on ROA in Go-Public Banks Listed on Indonesian Stock Exchange. http://eprints.undip.ac.id/28750.

[2] El Maude, J. G., Abdul-Rahman, A., and Ibrahim, M (2017) Determinants of Non Performing Loans in Nigeria's Deposit Money Banks. Archives of Business Research, Vol. 5(1), pp. 74–88. doi:10.14738/abr.51.2368.

[3] Hafiz, M. S., Radiman, Sari, M., and Jufrizen (2019) Analysis of Determinant Factors of Return On Asset in State-Owned Banks Listed on Indonesian Stock Exchange. Journal of Management and Financial. Vol. 8(2), pp. 107–122.

[4] Hermawan, Dwi. Fitria, Shoimatul (2019) Effect of CAR, NPF, FDR, and BOPO on Profitability Levels with Size Control Variables (Case Study at PT. Bank Muamalat Indonesia Period 2010 – 2017). Diponegoro Journal of Management. Vol. 8(1), pp. 59–68 http://ejournal-s1.undip.ac.id/index.php/djom. ISSN (Online): 2337–3792.

[5] Migustin, Puspa Erika (2017) Analysis of Factors Affecting BOPO in Islamic Banking in Indonesia. E-journal: repository.trisakti.ac.id

[6] Nahrawi, Amirah Ahmad (2017) The Influence of Capital Adequacy Ratio (CAR), Return On Assets (ROA) and Non Performing Financing (NPF) on Murabahah BNI Syariah Financing. Journal of Perisai. Vol. 1(2), pp. 59–98. doi: http://doi.org/ 10.21070/perisai.v1i2.881

[7] Nanda AS, Hasan, AF and Aristyanto E (2019) The Effect of CAR and BOPO Against ROA in Islamic Banking in 2011-2018. Perisai: Islamic Banking and Finance Journal. Vol. 3(1). doi: 10.21070/perisai.v3i1.2160

[8] Ozili, P. K (2017) Bank Profitability and Capital Regulation, Evidence from Listed and non-Listed Banks in Africa. Journal of African Business. Vol. 18, pp. 1–27. doi:10.1080/15228916.2017.1247329

[9] Puspitasari, Diana (2009) The Effect of CAR, NPL, NOP, NIM, BOPO, LDR and SBI Interest Rate Against ROA. Diponegoro Journal of Management. http://eprints.undip.ac.id.

[10] Poernamawatie, Fahmi (2009) The Effect of Credit Risk on Financial Performace of State-Owned Banks listed on the IDX. Journal of Gajana Management. Vol. 6(1), pp. 71–90.

[11] Rizkika, Refi, Khairunnisa,Vaya Dillak, Juliana (2017) Analysis of Factors Affecting the Profitability of Islamic Commercial Banks in Indonesia (Study on Sharia Commercial Banks Registered in the Financial Services Authority during 2012–2015). e-Proceeding of Management. Vol. 4(3), pp. 2675. ISSN: 2355–9357.

[12] Ratnawaty, Marginingsih (2018) Factors Affecting the Profitability of Islamic Commercial Banks in Indonesia. Journal of Ecodemica, Vol. 2(1).

[13] Riyadi, Slamet. Yulianto, Agung (2014) The Effect of Results Financing, Financing of Selling-Purchase, Financing to Deposit Ratio, Non Performing Financing on Profitability of Islamic Commercial Banks in Indonesia. Accounting Analysis Journal. Vol. 3(4), pp. 466–474.

[14] Tarawneh, A., Khalaf, B. K. A., & Assaf, G. A (2017) Noninterest Income and Financial Performance at Jordanian Banks. International Journal of Financial Research. Vol. 8(1), pp. 166–171. doi:10.5430/ijfr.v8n1p166.

[15] Teshome, E., Debela, K., and Sultan, M (2018) Determinant of Financial Performance of Commercial Banks in Ethiopia, Special Emphasis on Private Commercial Banks. African Journal of Business Management. Vol. 12(1). doi:10.5897/AJBM2017.8470.

[16] Tristiningtyas, Vita and Mutaher, Osmad, Drs M.Si (2013) Analysis of Factors Affecting Financial Performance in Islamic Commercial Banks in Indonesia. Indonesian Journal of Accounting. Vol. 3(2), pp. 131–145.

[17] Ubaidillah (2016) Analysis of Factors that Affect the Profitability of Islamic Banks in Indonesia. El-JIZYA Islamic Economics Journal. Vol. 4(1). ISSN 2354-905X.

[18] Wilantini, F, Dewi, NTB (2019) Islamic Finance Challenge in the Era of Four Point Zero. Perisai: Islamic Banking and Finance Journal. Vol. 3(2). doi: 10.21070/perisai.v3i2.2627.

[19] Yunita, Rima (2014) Factors Affecting The Level of Profitability of Islamic Banking in Indonesia (Case Studies on Islamic Commercial Banks in Indonesia, 2009–2012). Indonesian Journal of Accounting, Vol. 3(2), pp. 143–160.

Innovative and Intelligent Technology-Based Services for Smart
Environments – Ben Slama et al (eds)
© 2021 Taylor & Francis Group, London, ISBN 978-1-032-02030-3

Author index